Making things from
DISCARDS

Making things from DISCARDS

Beautiful and Practical Creations
with Bottles, Bread Dough, Tin Cans, Egg Cartons,
Plastics, and Folded Magazines

by Hazel Pearson Williams

BOUNTY BOOKS · NEW YORK

CONTENTS

INTRODUCTION

Housewives of every social stratum and every age group have already discovered the cleanest, most productive remedy for chasing blues and boredom out the door when leisure time hangs heavy—the remedy is handicrafts, and the results are invariably rewarding.

You might wonder, "What could be more exciting than making something beautiful with your own hands?" The challenging answer is "Making something useful or beautiful out of an object one normally throws into the wastebasket!" It's called "scrap handicraft," and it works like magic: tin cans become doll furniture; plastic bottles become dolls; stale bread is transformed into flowers and figures; and egg cartons are turned into fabulous flowers, wall ornaments, and whimsical creatures. And the "abracadabra" is all in your very own hands!

And, of course, since the cost is "zilch" to the maker, scrap handicraft can be enjoyed by many people who might otherwise be stymied by budget limitations— people like retired folks, den mothers, public and private school teachers—you—or perhaps your own daughters or nieces.

If you belong to a club group, there is a good chance that the program director will "love you forever" if you offer to lead a "Make It and Take It" scrapcraft workshop for her. If you have the good fortune and zeal to undertake such a project, make sure you have extra throwaways on hand for members who come unprepared. And do encourage the ladies to get in the act by saving their own throwaways and bringing them along with their scissors to future meetings. They *will* be coming back for more!

For those of you who may be timid about starting a throwaway project by yourselves, let me remind you that there are handicraft departments in many large department stores—and thousands of small specialty shops throughout the United States. Check the yellow pages for one in your neighborhood. They'll have the paints, glue, jewels, and glitter you may need to finish your project. They may also have classes—at nominal fees—where you can cozy with other "throwaway" cronies and decorate to your heart's content.

Have fun—
Hazel Pearson Williams

About the Author

Hazel Pearson Williams, president of Hazel Pearson Handicrafts, has attained stature nationwide as a prolific innovator of craft ideas. Her twenty-six years in the handicrafts business, coupled with her reputation as an inspiring craftswoman, have earned her a seat on the Board of Directors of the Hobby Industry Association of America. Mrs. Williams serves as Craft Consultant to Craft Course Publishers, a company that has published over a hundred "how to" crafts books for the consumer market.

In real life Hazel Pearson Williams is Mrs. Jack K. Williams and is the mother of two daughters and grandma of three grandsons as well. During her many years in the handicrafts business, she has become known as the "Queen of Crafts." She is president of Hazel Pearson Handicrafts, an international handicrafts supplier, with offices in New York City and Rosemead, California.

ACKNOWLEDGEMENTS

I would like to give credit to the following persons for their inspiration and help in making this book possible:
Roberta Harmon Raffaelli, Author, Creator, Designer of Magic with Tin Cans.
Amy Theisen, mother of ten energetic youngsters and clever crafts lady, for the Egg Carton Magic.
Ana DuPont, as beautiful and artistic as her name, for Bread Dough Artistry.
Maude Savage, lovely mother and grandmother who has "ideas that reach up to the rainbow," for Egg Carton Flowers.

H. P. W.

Bottle Cutting and Decorating

Fig. D Polishing accessories for power tools.

Fig. E Applying epoxy to section of glass.

NOTE: There are three projects in this book where the cut edges should be ground perfectly smooth. The others require only that sharp edges be removed. If polishing these projects seems tedious, your local glass man will gladly polish them for a nominal charge.

The two mandrels and the emery cloth shown in Fig. D, fit a ¼'' electric drill and are recommended for the hobbyist with a workshop. Power polishing will reduce the time and work of hand polishing. The small belt is used to smooth the inner and outer edges while the major grinding is done on the disc. WARNING — mount the drill in a permanent fixture on a bench. DO NOT hold the drill while grinding. Bench mounted wet-type belt sander is recommended for the person who makes bottle decorating a major leisure activity.

ASSEMBLING THE GLASS SECTIONS:

Assembling and gluing glass sections together to form a piece of original glassware is always a satisfying and rewarding experience.

Either a 5-minute epoxy glue or a flexible silicone adhesive is excellent for bonding glass. Each will cure to a permanent bond overnight, is waterproof and withstand age well.

The epoxy comes in two tubes — one being the resin and the other being the hardener. These are mixed in equal portions to form a soft, easy-spreadable adhesive. See Fig. E. Silicone is a rubbery adhesive which is squeezed directly from the tube to the area being glued. Any excess can be trimmed away with a razor blade after the adhesive has dried. Glass sections should be clean and dry before applying the adhesive. Allow the bond to cure before moving the project. □

Relish Dish

Trio in trim! Chubby amber-colored root beer bottles are cut-down-to-size and fancied-up with gold-colored braid. A glass handle makes carrying easy. A perfect server for spicy condiments or relishes.

MATERIALS
3 Root beer bottles, amber
Polishing materials
Silicone or epoxy glue
Optional: Decorative trim,
   of your choice

DIRECTIONS
1. Cut the three bottles 2¾'' above the heel.
2. Smooth the edges of the bottom portions.
3. Place the three pieces together in a triangle. Roughen the surface where the bottles touch with emery paper and coat with epoxy. Place a rubber band around the three bottles to hold together while the glue dries.
4. The handle is formed of a 1'' wide glass ring. Cut the ring and sand the edges smooth. Glue the ring in an upright position where the three bottles meet in the center.
5. Add decorative trim of your choice. □

Bottle Cutting
and Decorating

This genie

.............knows it'll be love at first sight when you see your empty glass bottles transformed into a vintage collection of glassware beauties! This is a very good year for your designing imagination to indulge in new vistas of glass-cutting artistry. Have a splendid spree! Bottles are easily adaptable to new shapes and styles, no matter what the whims of your endeavors. First take a moment to scheme. Do you want to make a window decoration? Cut glass rings and hang them on an invisible cord. Perhaps you want a string of wind chimes — just assemble the tops of bottles on a cord.

What's more, you can pretty-up glass vessels with antique gold leaf paint or dabble in colored paints and make smart designs on your glass creations.

As for the tradition of taking-out the "empties" — that chore is a thing of the past and quite fortunately so for our modern ecology-minded society. The candle cups, goblets, glasses, vases, compotes and praline dishes you make will all be shining examples of your sincere efforts to forestall waste. They're perfectly stunning gifts, too!

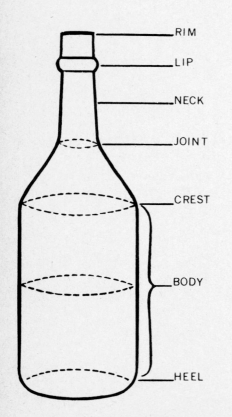

Let's Talk About Bottles

Bottles come in assorted shapes and sizes as well as many colors. Some bottles even have raised designs on them which are fun to work into a design on your project. Just remember to cut above or below the raised portion and not thru it.

So that we can talk about bottles more intelligently, general names have been given to the different parts as a point of reference. When in doubt, see the illustration at the left.

Bottle cutting is an art so take your time, work carefully and if you make a mistake, there are plenty of other bottles available. It is recommended that you read the general instructions on page 4 before beginning any project. □

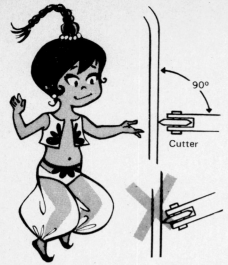

The Genie says to be sure to hold the cutter at a 90 degree angle to the bottle.

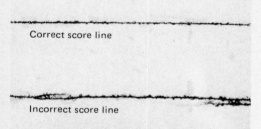

Correct score line

Incorrect score line

Fig. A Retouched picture of score lines.

Fig. B Rub inner and outer edge of glass with a piece of emery cloth.

Fig. C Polish cut edge by rubbing bottle in an emery slurry.

General Cutting & Gluing Information

Cutting and decorating glass bottles can be a most entertaining and satisfying family hobby. In many families the husband finds pleasure in cutting and polishing the bottles while the wife has the fun of assembling and decorating them.

When cutting and polishing a bottle it is important to consider the physical characteristics of glass. Glass is an extremely hard material, hence it is difficult to polish off the uneven edges of a poorly cut bottle. In the following copy we will show you how to keep glass polishing at a minimum by cutting bottles with a smooth, straight edge.

CUTTING THE BOTTLE:

Cutting a bottle requires two operations: *scoring* and *separating*. A bottle is scored by pressing a tiny rolling metal disc, called a "glass cutter", against the surface of the bottle. It is essential that the score line continue around the bottle and join the beginning; a clicking sound is heard when the ends meet. There are many excellent Bottle Cutters on the market that will accomplish this if you follow the manufacturer's instructions. It is also essential that the score line be a hair-like line rather than an interrupted, deep line with 'crackled' sides. See Fig. A. Here the manufacturer must rely on the hobbyist for good results since the critical factor is the force applied against the cutter. Other factors — dull or flat place on cutter; cutter not turning, cutter needs oil; cutter not held at 90 degree angle, see Genie; or scoring is too rapid ('burning' the glass). We urge you to practice making a good score line by making multiple lines every ½" on one or two clean bottles. Practice until you recognize a good score line by the sound the cutter makes. If you hear nothing, apply more pressure; if there is a grating sound, apply less. When the correct pressure is applied it will sound like tearing cotton fabric. Practice faithfully, for unless the bottle is correctly scored, the cut will be uneven and difficult to polish smooth.

SEPARATING THE BOTTLE:

Separating the bottle at the score line can be done by either of two popular methods. A bottle can be separated by gently tapping around the inside of the bottle at the score line — (a tapper is supplied by some manufacturers) — or by heating the full length of the score line. A variation of this method is to alternately heat and chill the outside of the glass at the score line. The source of heat can be either flame or electric and the chill can be obtained with a damp cloth or a piece of ice. With either method, cover the work surface with newspaper, wear cotton gloves and protect your eyes with glasses.

SMOOTHING THE CUT EDGE:

The sharp edges of a cut bottle are ground down with silicon carbide, available in hardware and craft departments as a loose grit or as emery cloth. Always work on newspaper to gather any loose grit and frequently dip the edge of the bottle in water to keep powdered glass from floating in the air. Double a 1½"x 6" piece of No. 120 emery cloth and rub it against both inner and outer edges until smooth. See Fig. B. The major polishing is done as illustrated in Fig. C. Place a 9"x 12" piece of glass on the work surface and pour a teaspoon of coarse grit onto it. Add a drop of water to make a slurry and grind the edge by moving the bottle in a circular motion.

Fig. D Polishing accessories for power tools.

Fig. E Applying epoxy to section of glass.

NOTE: There are three projects in this book where the cut edges should be ground perfectly smooth. The others require only that sharp edges be removed. If polishing these projects seems tedious, your local glass man will gladly polish them for a nominal charge.

The two mandrels and the emery cloth shown in Fig. D, fit a ¼" electric drill and are recommended for the hobbyist with a workshop. Power polishing will reduce the time and work of hand polishing. The small belt is used to smooth the inner and outer edges while the major grinding is done on the disc. WARNING — mount the drill in a permanent fixture on a bench. DO NOT hold the drill while grinding. Bench mounted wet-type belt sander is recommended for the person who makes bottle decorating a major leisure activity.

ASSEMBLING THE GLASS SECTIONS:

Assembling and gluing glass sections together to form a piece of original glassware is always a satisfying and rewarding experience.

Either a 5-minute epoxy glue or a flexible silicone adhesive is excellent for bonding glass. Each will cure to a permanent bond overnight, is waterproof and withstand age well.

The epoxy comes in two tubes — one being the resin and the other being the hardener. These are mixed in equal portions to form a soft, easy-spreadable adhesive. See Fig. E. Silicone is a rubbery adhesive which is squeezed directly from the tube to the area being glued. Any excess can be trimmed away with a razor blade after the adhesive has dried. Glass sections should be clean and dry before applying the adhesive. Allow the bond to cure before moving the project. □

Relish Dish

Trio in trim! Chubby amber-colored root beer bottles are cut-down-to-size and fancied-up with gold-colored braid. A glass handle makes carrying easy. A perfect server for spicy condiments or relishes.

MATERIALS
3 Root beer bottles, amber
Polishing materials
Silicone or epoxy glue
Optional: Decorative trim,
    of your choice

DIRECTIONS
1. Cut the three bottles 2¾" above the heel.
2. Smooth the edges of the bottom portions.
3. Place the three pieces together in a triangle. Roughen the surface where the bottles touch with emery paper and coat with epoxy. Place a rubber band around the three bottles to hold together while the glue dries.
4. The handle is formed of a 1" wide glass ring. Cut the ring and sand the edges smooth. Glue the ring in an upright position where the three bottles meet in the center.
5. Add decorative trim of your choice. □

Glassware . . . .

from Basic Cuts

CHEERS!
Serve your friends from your own distinctive glassware, and they'll toast your creativity! Bold horizontals and verticals combine to compose original looking forms — glasses just right for any beverage or vases for floral bouquets. Fun to make and to use!

1
2
3
4

The four pieces of glassware shown here were made from two kinds of beer bottles, Michelob and Olympia. Truly the variety of designs that can be made by combining pieces of different bottles is endless. Note that altho the vase and goblet appear quite different, they are alike except for the location of the cut made in the Michelob bottle.

Don't let any project limit your imagination — Try ideas of your own for that original look!

DIRECTIONS

1. FLOWER VASE:
   Cut one Michelob and one Olympia bottle 1" above the heel. Glue the rim of the Michelob to the Olympia bottom.

2. JUICE GLASS:
   Cut a Michelob bottle 3" up the body from the heel. Completely finish the edge.

3. GOBLET:
   Cut an Olympia bottle 3" up the body from the heel. Completely finish the edges. Glue the bottom section of the bottle onto the top section.

4. TUMBLER:
   Cut a Michelob bottle at the crest. Cut an Olympia bottle 1" above the heel. Completely finish the edges. Glue the rim of the Michelob to the bottom of the Olympia. □

PATIO BELLS . . . . . .
The soft, merry tinkle of glass bells in a gentle breeze dramatizes the beauty of any patio and lures the listener to "seek and spy" where the sound is coming from. The tops of bottles, in cool greens or warm browns, are easily fashioned into musical delights. The "know-how" for the patio bells can be found on page 9.

FLICKER GLOW . . . . .
Our "light show" glows on! Everyone loves the warm glow of candlelight — be it in your living room, dining room or den, these Gothic-looking, stained glass candle holders readily become status symbols for your home decor. The vivid colored glass is given an antique look with gold leaf trim. You can be proud of their radiant elegance! Find out how easy they are to make — see page 16 for the instructions.

Bring a touch of outdoors inside with a bottle terrarium. Turn to page 22 for instructions.

A colorful glass stain design enhances this candy dish. Know-how can be found on page 12.

A brown glass bottle candle-tainer is nestled in the midst of a bright colored Del Robbia wreath for a festive touch. Complete instructions are on page 10.

7

Fruit Compote

MATERIALS

Grand Marnier bottle or 4¼" dia. bottie
Fruit of your choice
Epoxy glue   Polishing material

DIRECTIONS

1. Cut the bottle at the crest and again 2¾" up from the heel. Discard the middle section.

2. Smooth cut edges .

3. Glue the rim to the center of the bottom of the heel. The top of the bottle now forms the graceful foot of the compote while the lower portion of the bottle forms the bowl. □

Planter

MATERIALS

1½ qt. Bottle, green   Polishing material
Epoxy glue   Planter mix
                            Small indoor plants
                            Gold braid

DIRECTIONS

1. Cut the bottle at the crest and smooth the cut edges.

2. Scratch the rim and center of the bottom of the heel with emery cloth and glue together with epoxy. Allow to dry.

3. Optional: Glue gold braid around bottle.

4. Fill the planter to within 1½" to 2" of the top with dampened planter mix and plant the foliage. □

Wind Bells

For whom do the bells toll? For all who have an open ear for graceful tinkling music! "Wind Bells"' glass bottle tops compose a natural melody as they sway in outdoor breezes. After a hard day's work, who wouldn't welcome a soothing, melodic sonata? Our bottle orchestra is ready to perform whenever nature's winds blow. "Wind Bells" aren't only colorful music for the ear — they're a delight for the eyes, too. Choose bright colored bottles; you may even find some with embossed designs. Gold-colored cord and a special tassel or bead-type clapper make the "connection". What a sound decorating accent for patios!

MATERIALS
5 Beer bottles
1 Yd. gold cord
3½" x 3½" Brass, paper-thin
Wire coat hanger
Polishing material

DIRECTIONS

1. Cut each of the five bottles at the crest. Only the tops will be used for bells. Sand the cut edges to remove sharp edges.

2. Cut the wire coat hanger into five 3" lengths. Your pliers may also cut wire.

3. Bend each wire into a hair-pin shaped hook. See Fig. A.

4. Cut the brass into a triangle clapper. Punch a hole in the top and tie to one end of the gold cord.

5. Tie a wire hook onto the cord 2" above the brass clapper.

6. Thread the first bell along the cord until it rests on the wire hook.

7. Tie another hook onto the cord 2" above the previous one; add another glass bell.

8. Continue until five bells are in place.

9. Double the remaining cord to use as a hanger for the wind chimes.

10. Optional: Large beads may be added to the cord above the bells, if desired.

The directions given are for the tinkling bells shown at the right. The other bells are made using the same principle.

Budweiser beer bottles were used for the center wind bells while three Carlsberg beer bottles were used for the bells on the left.

Fig. A Bend wire into a hair-pin shape.

Del Robbia Candletainer

Fall holiday time means "turkey talk" and decorations! Here's a fine, easy-to-make expression for your dining table — a charming Del Robbia Candletainer. A tall center candle, encircled by a bottle, lends a romantic accent to the occasion. Why not choose a scented candle to enhance the atmosphere with delicious fragrance? The rich colored fruit, in the wreath, create a festive arrangement that's just right for displaying in a glass bowl.

MATERIALS

Footed amber glass bowl of your choice
Wine bottle, amber, 1/5 size
Del Robbia wreath
Candle, 2" to 3" dia., harvest gold color
Floral clay

DIRECTIONS

1. Cut the bottle at the crest; discard the top portion and smooth the edge of the bottom portion.

2. Secure the cut bottle to the bottom of the bowl with floral clay. This will become a candletainer to hold the candle.

3. Arrange the Del Robbia wreath around the bottle as shown in the photo.

4. Secure the candle in the candletainer with floral clay. If needed, the candle can be cut off at the bottom so it doesn't extend beyond the bottle. ☐

Christmas Ornament

Peek-a-boo! Tiny imported novelty figures insure a happy holiday when used in glass ornaments. You'll be going in circles trying to decide where to hang them. These glass "cut-ups" spread cheer hanging from windows and mirrors, as well as on your Christmas tree.

DIRECTIONS

1. Cut a 1" wide glass ring and smooth the edges.

2. Use tacky craft glue and trim one edge of the ring with strung sequins and the other edge with decorative gold braid.

3. Glue a small Santa or Christmas figure of your choice to the inside bottom of the ring. Allow the glue to dry.

4. The ornament is ready to suspend from a length of monofilament or nylon cord. ☐

Patio Light

"No deposit, no return" — except in extra special lighting effects from these glass bottles. Light up the game room, bar or patio with this bold swag lamp. Seven soda bottles are painted frosty yellow, pink, blue, and green with Cryst-l-craze — a touching circle for a ½ gallon wine bottle center.

MATERIALS

7 Soda bottles, clear
½ gal Wine bottle
Polishing material
Epoxy glue
Rubber bands, heavy
Lamp parts
Cryst-l-craze paint & brush

DIRECTIONS

1. Cut the heel off each of the seven bottles. Sand edges.

2. Cut the seven bottles at the joint and sand the edges.

3. Cut the heel off the ½ gallon bottle. Sand the edge.

4. Place the seven bottles in an upright position around the ½ gallon bottle. Hold them in position with heavy rubber bands. Adjust the position of the bottles so they are evenly spaced.

5. Mix epoxy; place some on the end of a screw driver and dab the glue between the center and outer bottles. Allow glue to dry.

6. Turn the lamp upside down and epoxy the bottles at the bottom. Allow the glue to dry.

7. Paint the outside of each bottle with a different color of Cryst-L-Craze. The places where the bottles are glued together and can't be reached with a brush, should be touched up with paint on the inside.

8. See page 13 for electrical wiring instructions. □

Decoupage Jars

An apothecary jar, decorated with a colored print, becomes a useful and decorative item for your home. Since the prints are coated with clear acrylic, they can be cleaned with a damp cloth without harming them.

The quaint prints, on the round jars, were cut in oval shapes to fit each jar. The square jars were decorated with cut-out mushroom prints.

Seal the prints with clear acrylic spray before cutting them out with manicure scissors. Adhere the prints to the jars with white craft glue. □

11

Stain Glass Candy Dish

A dish of irresistable mints to nibble is fun to have around. A stunning candy dish can be made by cutting and reassembling a coffee jar. Follow the easy directions below for decorating it with gold paint and glass stain.

DIRECTIONS

1. Clean the glass with a solution of white vinegar and water for good adhesion.

2. Draw a simple design, of your choice, on a piece of paper. Tape it to the inside of the bottle. Outline the design with gold paint.

3. To apply glass stain colors, hold the glass surface horizontally. Load a soft bristle sable brush with stain and flow it onto the glass with as little pressure on the brush as possible. This will give an even coat without brush marks. Two or three coats can be painted on for a more intense color. Allow ½ hour drying time between each coat.

4. Cryst-L-Craze is a transparent glass stain paint which dries to an interesting crackled effect. Apply just one heavy coat and watch it crackle as it dries.

5. Strips of ¼" lead can be glued around the top and bottom of the candy dish for a decorative trim. □

Leaded Glass Candletainer

When lead strips are adhered to the surface of a bottle and it is painted with glass stain, a beautiful effect is achieved. ¼" lead stripping and contact adhesive are available in most craft departments.

DIRECTIONS

1. Draw a simple design on a piece of paper and tape it to the inside of a cut bottle.

2. To apply lead strips to the outside of the bottle, cut the lead with scissors or a razor into 12" lengths. Straighten the lead by holding both ends of a strip and pull it over the edge of a table. Apply a thin coat of contact adhesive to the flat side; allow it to get tacky and press the lead to the glass following the design lines. If making curves or circles, shape them before applying the adhesive.

3. Tightly press the lead against the glass.

4. Paint the leaded sections with glass stain. If some paint gets onto the lead, wipe off with a cotton swab dipped in paint thinner.

5. Place a votive candle inside the container. □

Garden Light

This garden light is a truly exciting innovation for the garden lover. Simply select a piece of statuary which expresses the mood of your garden — wire it for electricity and top it with a cut bottle decorated with lead strips and painted with glass stain. You can make this attractive piece for a fraction of what a garden light would cost and think of all the fun you'll have doing it!

MATERIALS

Bottle, 1 gallon, clear
Lead strip, ¼" wide & cement
Glass stain paint & brush
Plaster garden statuary, of your choice,
    wired for lighting. Approx. 16" high
Madras tissue paper
White craft glue
Polishing material

DIRECTIONS

1. Cut the bottle at the crest and discard the top portion.

2. Sand any sharp edges just enough so they won't cut.

3. To make a pattern for the lead strips, cut a piece of shelf paper 4½" wide, to fit the inside circumference of the bottle. Fold the paper in half and then in half again. Pencil in these fold lines and then draw a X thru each section.

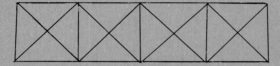

4. Tape the pattern to the inside of the bottle.

5. Cement the lead strips to the outside of the bottle and allow the cement to dry.

6. Color the inside of the bottle with glass stain paint.

7. Glue a 4½" wide strip of Madras tissue paper around the inside of the bottle. This gives an interesting, opaque color effect and conceals the bulb and socket. □

HERE'S HOW TO WIRE A BOTTLE LAMP SUCH AS SHOWN ON PAGE 11

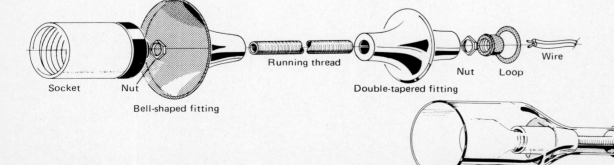

Socket   Nut   Running thread   Wire

Bell-shaped fitting   Double-tapered fitting   Nut   Loop

BRIGHT NIGHTS BEGIN WITH BEAUTIFUL LIGHTS . . . . .

The lamp shown at the upper left, was designed to inspire creativity and is shown as an idea for using your ½ gallon jugs and old lamp parts. Information on wiring a bottle fixture is found on page 13.

Suspend a little splendor by hanging the lamp shown at the lower right, in the patio, over a game table or bar. Instructions can be found on page 11.

Inspired illumination for garden decor! The garden light shown at the lower left, is easily made by placing a section of a bottle on top of a piece of plaster statuary. Turn to page 13 for the instructions.

Rescue windows from the "doldrums" with Window Charmers —
like ones trimmed with quaint wooden figures or rice paper with
dried flowers embedded in it. Turn to page 21 for instructions.

GLASS DOME FOR DISPLAYING DRIED FLOWERS

*Expensive to buy ------
but fun and inexpensive to make!*

Cut a ½ gallon bottle at the crest and discard the top section. Smooth the cut edge of the glass dome.

Cut a 4″ and 6″ dia. disc from a 1″ thick sheet of insulation cork. Glue the 6″ disc to the top of the 4″ disc to form the pedestal.

Attach a small ball of floral clay to the center of the pedestal and arrange the stems of dried flowers and pods in it.

Invert the glass dome over the floral arrangement.

Flicker Glow

All cracked-up and amazingly beautiful! The look of stained glass is borrowed from yesterday, but the unusual combination of colored glass and candlelight is definitely in step with the designs chosen by young moderns. Firey candlelight diffuses through fractured glass with a crystal-like sparkle. Liquid gold leaf brush strokes add rich outline borders to each colored section. Larger containers hold two or three votive candles — intensifying the exciting lighting effect.

MATERIALS

½ Gallon bottle or size desired
Tempered glass, untinted (the
    back window of an automobile)
1'' Brush
No. 1 Round, sable brush
White craft glue
Stain glass paint, red, blue,
    yellow and green
Liquid gold paint
Plaster of Paris   Cotton swabs
Votive candle   Acetone

DIRECTIONS

1. Place the clean tempered glass window on newspaper or a painter's drop cloth. Paint glass stain on the window in four different sections: red, blue, green, and yellow. Apply a liberal application of paint with a 1'' brush. Allowing the paint to flow from the brush onto the glass gives a good coverage and produces colors of bright intensity; light, washed-out colors are not desirable. Allow the paint to dry thoroughly.

2. Completely wrap the painted glass in an old sheet and place on a hard concrete surface such as a driveway, walk or patio. Fracture the glass by striking the covered window ONCE with a hammer.

3. Prepare a box for storing each color of glass. Carefully lift the shattered glass into the box with a spatula to prevent breaking it into smaller pieces.

4. Cut a ½ gallon bottle 5½'' up the body from the heel. Discard the top portion. Sand the cut edge to a smooth finish.

5. With a 1'' brush, coat the entire outside surface of the bottle with white craft glue. Allow the glue to dry.

6. Paint a 1'' wide strip of glue vertically down the bottle. Place a few long, narrow pieces of fractured glass into the damp glue. See Fig. A. Press the glass firmly against the side of the bottle. Allow the glue to completely dry. This strip of glass is important since it will serve as an anchor for the other pieces of glass to be pushed against in the next step.

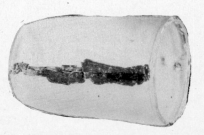

Fig. A Glue a long, narrow strip of glass to the bottle.

Fig. B Apply the plaster grout.

7. Cover the bottle with irregular pieces of fractured, colored glass, as for a mosaic. To do this apply some glue to the bottle, select a piece of glass, the color and shape you desire, and press it into the glue; push snug against the anchor strip. Be sure the glass is flush against the bottle so that the plaster grout, applied in step 8, will NOT seep between the glass and the bottle. Continue until the bottle is covered.

8. Mix the plaster of Paris with water to the consistency of mashed potatoes. Wearing rubber gloves, pat the plaster grout into the cracks and crevices around the fractured glass. See Fig. B.

9. Quickly remove excess plaster by wiping the glass with a damp turkish towel. See Fig. C. Allow the plaster to dry.

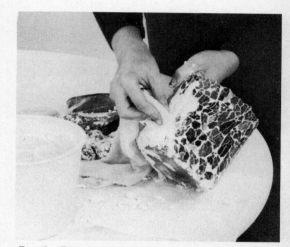

Fig. C Remove excess grout with a damp towel.

10. Turn the decorated bottle upside down and coat the bottom of the heel with plaster. Be sure the plaster is even so the bottle will set straight. Print your initials and date in the damp plaster with a toothpick, pencil or pen knife. Allow to dry before turning upright.

11. Plaster the top rim of the bottle. Wipe any excess plaster off the glass and allow to dry.

12. Apply gold paint to the plaster grout with a No. 2 sable brush. See Fig. D. Try not to get any gold on the colored glass. If gold does get on the glass, remove it with a cotton swab dipped in acetone. Gold paint the plaster around the top rim. Allow the gold to dry.

13. Invert the decorated bottle and apply gold to the plaster on the bottom. Allow to dry.

14. The Flicker Glow is ready to use by placing one, two or even three votive candles inside it. □

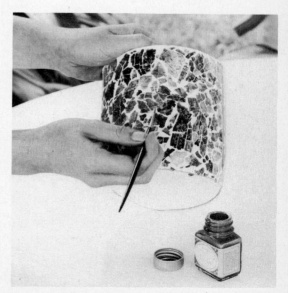

Fig. D Paint grout with gold paint.

Herb Garden

Parsley, sage, rosemary, and thyme — win your true love's heart with a little spice at a time! Who could resist your special recipes when you prepare them with fresh grown, natural spices — right from your own Herb Garden. Palates will tingle with the excitement of full-bodied flavors and rich aromas. Four green champagne bottles are used as containers for the herb assortment of your choosing. Each bottle is encircled with an attractive copper strip studded with cap nuts. Arrange the bottles in a row of four, as shown above, or two by two, forming a square. Besides being delectable, this Herb Garden makes a living showpiece for window shelves.

MATERIALS
4 Champagne bottles, 1/5 size
8 Bands of copper, ¾''x 12'' each
Polishing material
Epoxy glue
40 Cap nuts, ¼''
Masking tape
Pebbles and planter mix
4 Herb plants of your choice

DIRECTIONS
1. Cut each bottle 5'' up the body from the heel and discard the tops. Sand the cut edges to a smooth finish.

2. Cut two copper bands to exactly fit around each bottle. Epoxy glue one a ½'' up from the heel and the other ½'' down from the top. Tape to hold until the glue is set.

3. As a decorative trim glue five cap nuts to each copper band, at even intervals. See photo above.

4. To plant, place pebbles in the bottom of each bottle and add planter mix. Plant a different herb in each bottle, being sure to press the roots in firmly. We used: Lemon Balm, Black Sage, Thyme, and Sweet Basil.

5. Set your bottles side by side in a sunny or semi-sunny window so the plants will get the sunlight they need for growth. □

Wind Chimes

Even the most meticulous have some odd 'n ends of broken bottles. Don't despair! Face the music and create a wind chime for hanging in the patio or garden.

DIRECTIONS

1. Smooth sharp edges off glass with polishing materials.

2. To prepare each piece of glass for hanging, twist a small loop in the center of a 2½'' length of fine wire. Bend each end of the wire into a flat coil, as shown at right. Glue a coil to each side of the glass. Conceal coils by gluing a small bead or shell over each.

3. Heat an awl until hot enough to melt a hole in the plastic dish. Space holes about 4'' apart.

4. Use different lengths of monofilament cord for hanging the glass. Thread one end of the cord thru the hole in the plate and secure to the top by tying on a bead. String 3 or 4 beads on the other end of the cord and slip them up into place. Secure the bottom bead with glue. Add a piece of glass by tying thru the wire loop. To hang another piece of glass, glue a wire loop to the bottom of the first piece of glass, string on another piece of cord, add beads and then add another piece of glass. See figure at the right.

5. Continue hanging glass pieces so they will touch and tinkle when moved by a breeze.

6. Add a cord to the top for hanging the plate. □

MATERIALS

Assorted pieces of glass: broken bottles or stain glass
Oval plastic dish, approx. 9''x 18''
Cord for hanging wind chimes
Polishing material
Epoxy glue
Assorted beads
24 ga. Wire
Monofilament cord

Ship in a Bottle

MATERIALS
Bottle, one gallon
Cork stopper
Small model ship of your choice
Driftwood base, approx. 7" x 12"
2 yds. Decorative braid
Epoxy glue
Masking tape

DIRECTIONS
1. Cut the gallon bottle 2" up the body from the heel.

2. Lay the large portion of the bottle on its side. Insert a model ship, as shown in the photo above, and secure it with epoxy glue. Allow the glue to dry.

Now, the age old mystery of the sea comes to life again in an unique "ship in bottle" theme that is nautical, but nice — and easy to make! Handsomely designed to intrigue both seafarers and landlubbers alike. The unusual display is perfect for exhibiting in the study or den, or on a desk. Berthed inside the one-gallon sized bottle is a miniature replica of the famed "Pinta". Have the pleasure of making your own model ship — deck the "Pinta" in patriotic red, white, and blue and set her sails at full mast, or purchase a preassembled ship. Husky rope braid trims the bottle. An ideal gift for nautical enthusiasts. Come "sail" through the easy instructions.

Window Charmers

Charmers for distinguished circles only. Decorate glass circles with gingham, dried flowers, tiny novelties and hang them in your window. They're fun!

MATERIALS
2 Soda bottles, clear
Polishing material
Epoxy glue
Mylar
Monofilament or nylon cord
Decorations as desired
    Braids and cords
    Miniature figures
    Rice paper with embedments
    Glass stain paint

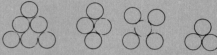

Fig. A Suggested designs for glass rings.

DIRECTIONS
1. Cut at least six 1'' glass rings from clear bottles. Sand the edges.
2. Place the rings on wax paper or aluminum foil and arrange them in any one of the designs shown in Fig. A.
3. Apply epoxy glue on the top edge where the rings touch. Allow the glue to dry.
4. Turn the design over and glue the other side where the rings touch each other. Allow the glue to dry.
5. Decorate the rings as desired and hang on monofilament cord.

    Here are instructions for the rings shown.

a. Glue miniature wooden figures and plastic greenery in rings. Trim the edges with gold cord. See the large photo above.

b. Cut mylar circles to fit the back of the rings. Cover the mylar with calico and glue onto the back of the rings. Trim edges with rickrack. See Fig. B.

c. Cut mylar backgrounds and paint with glass stain. Glue on rice paper with floral embedments; secure to the back of the rings. Trim edges with gold braid. See Fig. C.

d. Cut smaller diameter glass rings. Back them with calico and glue to the larger glass rings. Add a wooden figure and dried flowers. Trim edge with braid. See Fig. D. □

Fig. B Calico decoration.

Fig. C Rice paper decoration.

Fig. D Rings within rings.

Terrarium

Make your own greenhouse — perfect for living rooms, dens or patios! A miniature lush and leafy "glen" encompasses all the peaceful beauty of nature's outdoors in a "bottle environment".

CARING FOR A TERRARIUM

To maintain moisture in the terrarium, cork the planted bottle. If the glass becomes fogged, remove the cork for an hour or two until the moisture escapes. A small amount of condensation should be showing at all times; if not, add a small amount of water. Additional water may be required only a couple times a year.

MATERIALS
½ Gallon, grape juice bottle
Lead strip, ¼" width & cement
Epoxy glue
Planting mix with charcoal
Small plants of your choice
Masking tape
Cork for bottle

DIRECTIONS

1. Cut the bottle 4¾" above the heel. Place a piece of masking tape on both halves of the bottle before separating it so you can line up the cut portions later.
2. Clean and dry the two halves.
3. Put masking tape over the cut edge of the bottom portion, to prevent injury while planting.
4. Add water to the planter mix until it is evenly moistened and blend in some granulated charcoal. The charcoal is important because it helps maintain an odorless atmosphere. Place 2" of the mixture in the bottom of the bottle and plant the foliage. Tamp the mix around the roots to eliminate air pockets. Some good plants — Dracena, Neanthebelle Palm, Pepperoni, Chinese Evergreen and Baby Tears.
5. Remove the tape from the cut edge and carefully apply epoxy glue. Glue the bottle together, lining up the cut. Allow glue to dry.
6. Cement a strip of lead trim over the cut seam.
7. Allow the plants to acclimate for several days and then insert a cork into the top of the bottle.

Advanced hobbyists may want to cut and plant a five-gallon bottle. See the photo at the right. Most bottle cutters have special instructions for cutting large bottles. □

Egg Carton Magic

IDEAS FOR OVER 30 EGG CARTON NOVELTIES

Princess Angel

With the wave of a magic wand this Princess Angel steps from an egg carton fairyland right into your heart. She stands 6" high. Make her of white egg cartons for Christmas or soft pastels for Springtime. Can't you see her up and down the table for a Mother Daughter banquet?

MATERIALS

2 egg cartons, white
Small doll head, 1-1/2" dia.
Tinsel-tex wire, gold
Gold paper lace, narrow border
Clear all-purpose glue-glaze

DIRECTIONS

1. Cut three sections from the four cups on the ends of the egg carton. The flared bottom of cup should measure 2-3/4". See Fig. A.

2. Make the bottom tier of skirt by fusing these three pieces together from the inside. Pieces will overlap. See Fig. B.

3. Scallop bottom of skirt with cutter. Etch a lace design, of your choice, by burning through skirt.

4. Make a middle skirt tier the same as the first. Etch and then glue to first tier. See Fig. C.

5. Cut four 1-1/4" wide petal-shaped pieces from the side of the egg carton for the upper skirt tier. Etch lace design around the bottom edge. Glue one to front, back and each side of the second skirt tier. See Fig. D.

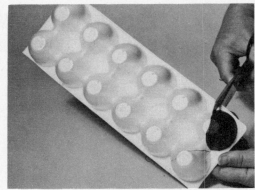

Fig. A Cut 1/3 of egg cup away.

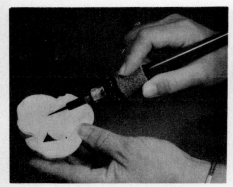

Fig. B Fuse first tier of skirt together on underside.

Fig. C Glue middle tier of skirt to first tier.

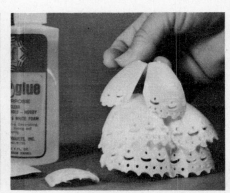

Fig. D Cut, etch and glue petal-shapes to middle tier.

6. Cut out a divider as shown in Fig. E. Etch a lace design and glue to upper tier of skirt to form a waist.

7. Cut out an egg cup and etch as shown in Fig. F for bodice and shoulders. Glue to waist.

Arm Pattern

8. Cut a hole in top of shoulders for head. Glue head in place.

9. Cut arms from curve on lid. See Fig. G. Use cutter to make arm holes and glue arms into place.

Wing Pattern

10. Cut wings from lid. Etch designs in wings. Trim opening at back of shoulders; insert and glue wings into place.

11. Twist a 5" length of gold tinsel-tex wire into a halo and trim with gold paper lace. Attach to head.

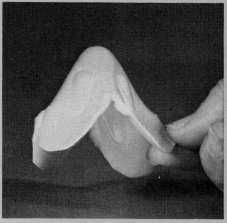

Fig. E Cut egg separator for third tier of skirt.

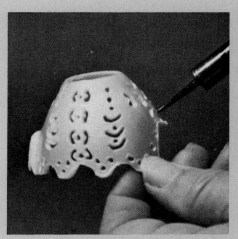

Fig. F Etch egg cup for bodice and shoulders.

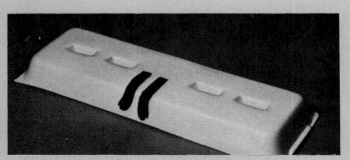

Fig. G Cut arms from curve on lid of carton.

A salvaged plastic egg carton is an inexpensive source of construction material for making clever craft novelties. The plastic is soft enough to be easily cut with scissors; can be readily fastened together with clear glue-glaze or tacky craft glue; colors can be changed to your heart's content, with tempera, all-purpose spray paint or acrylics. It must certainly become a most popular material for these reasons alone.

Yet the magical transformation from egg cartons to some of the novelties presented in this book depends on still another physical characteristic -- the unique reaction of the material when a heated tool, such as an awl or an electric cutter is placed against it. The material can be carved, fused together or cut off with heated tools. To heat the awl hold it over heat until the metal is well heated. You will find it necessary to re-heat the awl frequently. The constant tip temperature of an electric cutter makes it a desireable tool for use in egg carton craft.

For those of you who have a refrigerator full of eggs and still need additional cartons check with your craft dealer. Many store-keepers sell the plastic cartons without the eggs.

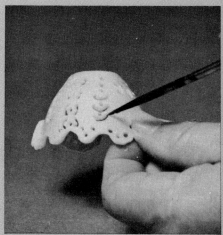

Heated awl is used to create a lace-like bodice.

Butterflies and Cattails

Butterflies, hovering on driftwood with make-believe cattails and colorful posies, make a stunning pièce de résistance for any coffee table or bookshelf. These attractive butterflies have a 3-1/2" wing spread.

MATERIALS

6 Egg carton lids
Clear all-purpose glue-glaze
Acrylic paints: black, brown yellow plus colors of your choice
Stem wire, 18 gauge covered
Bare wire, 24 gauge
Fine sandpaper
White craft glue
Driftwood

DIRECTIONS

1. WINGS: Make a master paper pattern. Cut a wing from the center of the carton lid. Pattern should overlap onto the curved side giving the wings a curved effect. See Fig. A. Be sure to turn pattern over to make a matched set. Cut decorative designs in wings, with a hot cutter, as desired. Make two pair.

2. BODY: Make master cardboard pattern. The body is made by stacking and pressing three pieces of lid together while cutting around pattern with a hot cutter. The pieces will fuse together as you cut. Trim edges for a rounded effect. Sand lightly to finish. Cut slits at an angle so wings will be angled. Make two bodies.

3. Paint butterfly wings a color of your choice. Paint bodies black or brown.

4. ANTENNA: Make two antennas for each body. See Fig. C. To make, curl one end of a 1-1/2" piece of 24 gauge wire around a small round object such as a pencil or paint brush. Push straight end of wire into position on head. Glue to hold.

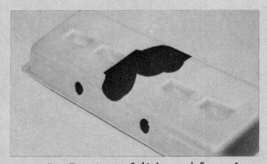

Fig. A. Portion of lid used for wing.

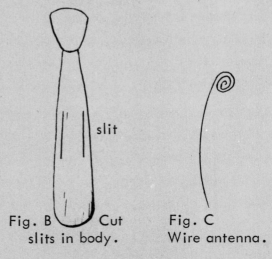

slit

Fig. B Cut slits in body.

Fig. C Wire antenna.

5. CATTAILS: Make a master cardboard pattern. For each cattail tightly press five lids together and cut around the pattern with a hot cutter. Trim round in the same manner as for butterfly bodies. Sand smooth. Make three cattails. Glue a stem wire into the bottom of each. Paint brown and dry-brush with yellow.

6. POSIES: Cut flowers, with stems and leaves, from a piece of lid. See Fig. D. Place cutouts on a cookie sheet and put in 300 degree oven for fifteen seconds or until flowers start to puff and curl. Remove from oven. Paint leaves and stems green and flowers in colors of your choice.

7. GRASS: Make grass by cutting pieces as shown in Fig. E. Follow the same procedure as for puffing flowers. Paint green.

8. ARRANGEMENT: Make small holes in a piece of driftwood, with a drill or nail. Position these where you want the cattails. Glue cattails and butterflies to driftwood. Arrange and glue flowers and grass into place. See colored photo above.

9. Brush coat all of the foam, except the cattails, with artist's gel or glue-glaze. This strengthens it and gives it a glossy finish.

Cattail
Pattern

Fig. D Pattern for posy with stem and leaves.

Fig. E Pattern for grass.

Candy Cup Favors

These candy cups are "eggs-actly" right for any fun-filled party! The egg cups are cut apart, trimmed into petal shapes, nested and glued together. Then they are decorated to suit your fancy. Use these ideas or create your own designs. Excellent for penny-wise banquet decor.

Fluffy, red chenille adds accent to these candy cups.

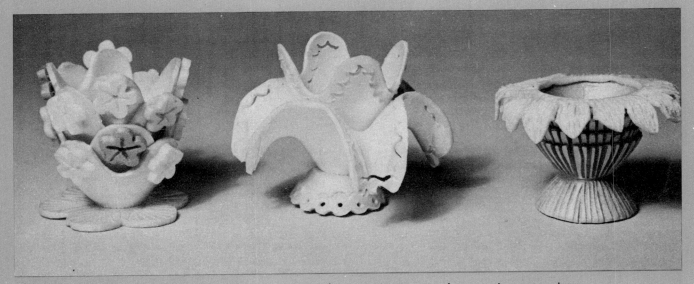

Notice the effective and unusual carved trim on the candy cups above and below. A hot cutter was used to carve the several designs.

True egg carton magic! A bowl full of yellow Shaggy Daisies is accented by a snow-white Princess Angel. See pages 24 and 25 for instructions.

Colorful egg cartons are a rich source of material for fascinating party favors and unusual candy cups.

Lovable Lamb

The realistic wool coat on this lovable 5" lamb was sculptured. It is one of many interesting effects that can be achieved with a heated awl or electric cutter. This same technique is used to complete other animals shown in color on pages 32 and 33.

MATERIALS
2 Egg cartons, white
Clear all-purpose glue-glaze
Pr. glass eyes, 8 mm
Acrylic paint, pink

DIRECTIONS

1. BODY: Cut and fuse two egg cups together. See Fig. A. Smooth edges with a hot cutter.

2. HEAD: Cut 3/4" piece from bottom of egg cup and 1" piece from volcano-shaped center divider. Fuse together. See Fig. B.

3. Fuse head to body by inserting hot cutter in hole at bottom of body and piercing, three times, where body and head touch. Head should be turned slightly to one side.

4. SHAPE: Slice a 1/4" curved section from volcano-shaped center divider and fuse around neck. Build up head by adding penny-size pieces cut from the egg cups.

5. WOOL: Cut penny-size pieces from the sides of the egg cups. They need not be round. Fuse circles, concave side down, one at a time, onto lamb. Overlap them and carve them for wool effect with the hot cutter. See Fig. C. Completely cover body, head and around face.

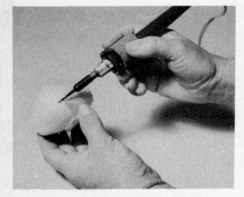

Fig. A   Fuse two egg cups together for lamb's body.

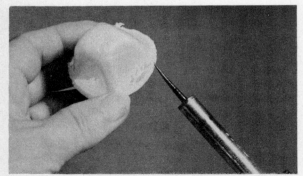

Fig. B   Fuse nose to head.

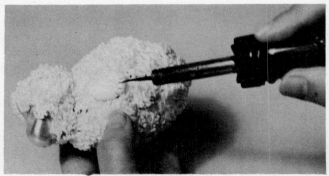

Fig. C   Fuse concave circles to body and carve with cutter for wool effect.

6. EARS AND TAIL: Follow patterns for ears and tail and cut pieces from corners of lid. See Fig. D. Use hot cutter to make a slit in head and body for ears and tail. Place in slits and glue in place with all-purpose glue-glaze.

7. LEGS: Cut out three 1"x 2-1/8" pieces from lid of carton. Stack the three pieces and fuse them together by cutting around the outside edge with a hot cutter. Hot cut lengthwise down the center to make two legs. Repeat process for other legs. Use cutter to round off edges of legs. Shape a hoof in end of each leg as shown in Fig. E.

8. Melt a hole in body for each leg and insert. Cut dime-size circles and fuse to top of legs. Fuse them to body.

9. Paint nose pink and allow to dry. Coat with all-purpose glue.

10. Glue glass eyes to head.

Ear Pattern   Tail Pattern

Fig. D Cut ears and tail from corners of carton lid.

Fig. E Shape a hoof in end of leg.

Carton Characters Are Fun

A basic armature can be made adaptable to many different kinds of characters. See photo above.

Simply make suitable clothing of felt, crepe paper or cloth. Add feet, ears, tail or what-have-you and presto --- you have an original!

To stabilize the armature put some floral clay inside the bottom egg cup for weight before fusing the two egg cups together for body.

The colorful menagerie made entirely from egg cartons is indeed magical. The jocular pig and the sad eyed lamb appear quite calm inside their egg carton corral. Neither the sitting lion nor the standing bear look very fierce anyway -- do they? The garland of egg carton flowers in the bear's hands are a pretty good indication of his temperament.

These animals are not nearly as difficult to make as they appear to be. The glamorous pig can be made in under an hour by

The other three fellows take longer because of their realistic fur coats but after you have made the lamb, the lion and bear are much the same. The egg carton armature is covered with penny-size pieces of egg carton which are carved into "fur" with a heated awl or an electric cutter. The oversize daisies and the corral fence only take a few minutes more.

Shaggy Daisy

A basket full of 4" Shaggy Daisies, in soft yellow, will brighten any corner. Makes a charming arrangement for patio, lanai, den or breakfast nook. See color photo on page 29.

MATERIALS
Egg carton, any color desired
Pep cluster
Daisy leaves
Stem wire, 18 gauge
Floral tape, green

1. Remove lid and lip of carton. Place the carton bottom up. Cut out four sections as shown.

2. Fringe each section by cutting down to the edge of the bottom of the cup.

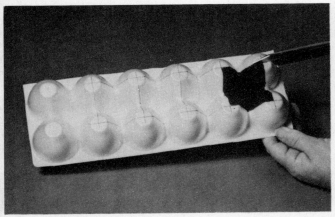

3. Place the four fringed sections inside each other. Rotate corners to form a good-shaped flower.

4. Fasten pep cluster to end of wire stem. Push through center of flower.

5. Add leaves while the stem is being taped.

Easter Lily

The majestic Easter Lily, in all its Springtime beauty, will capture the heart of all who see it. It is 4" in diameter. See page 36 for an arrangement of lilies in full color.

MATERIALS

Egg carton, white or yellow
6 Lily peps
Lily leaves
18 gauge stem wire
Acrylic paint, light green or
    chartreuse
Floral tape, green
White craft glue

DIRECTIONS

1. FLOWER CENTER: Make a cluster of six lily peps. Place cluster along stem wire. The peps should extend completely above wire. Secure peps to stem with floral tape. See Fig. A. Be sure to always pull stretch out of tape while wrapping stem.

2. With bottom side of egg carton up, cut four sections as in Fig. B. Each petal is made to look individual by cutting a "V" shape out of each cup. See Fig. C. Trim around outside edge, eliminating any sharp places.

3. ASSEMBLY: Push flower center and stem through center of petals. Apply glue to base of center to hold it in place.

4. LEAVES: Add lily leaves while taping down stem.

5. PAINTING: Dry-brush light green or chartreuse paint in the center of the lily for a finished effect.

TIGER LILY: Spots of acrylic paint may be added to petals to make a Tiger Lily. See Fig. D.

Fig. A Secure lily pips to stem with floral tape.

Fig. B Cut across cups for lily.

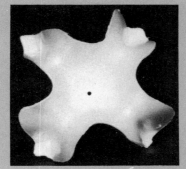

Fig. C Cut a "V" shape in the top of each cup.

Fig. D Add spots for a Tiger Lily.

EASTER LILIES are elegant in their simplicity. This regal flower is easy to make and lovely to use for Springtime decoration. Instructions are on page 35.

TIGER LILIES make a "fun" arrangement for the patio or lanai. It's a quick and simple step from Easter Lilies to Tiger Lilies. Just substitute yellow egg cartons and paint on brown spots. Magic for real!

POINSETTIAS --- in snow white and cheery red --- are featured in this lovely Christmas arrangement. Easy-to-follow instructions are found on page 39. Make this decoration for your holiday mantel or use it on the coffee table.

HOLIDAY TREE AND WREATH
This lovely setting certainly belies its humble beginning. The glamorous rosettes were cut from egg cartons and pinned to a foam wreath and a foam tree. The Madonna that graces the decoration is also handcrafted.

Holiday Tree

This festive Holiday Tree stands 14" high and is made of white egg cartons trimmed with pink parchment and gold medallions. The decorative rosettes can be made to cover foam wreaths, topiary trees or similar objects for a matched ensemble.

MATERIALS

6 egg cartons, color of your choice
1 pkg. 5-3/4" x 5-3/4" Pearl Parchment, color of your choice
30 Gold paper lace medallions, 1" dia.
8" White foam cone   30 Corsage pins
Bead Spray   White craft glue

DIRECTIONS

1. Cut a single cup section as shown in Fig. A.

2. Cut the sides to divide into eight parts. See Fig. B.

3. Cut the ends of each part into points. See Fig. C. Repeat for sixty cups.

4. Cut a sheet of Pearl Parchment into four squares. Fold each square into fourths and cut as shown in Fig. D. Open into an eight-petaled shape. Make thirty.

5. ASSEMBLING ROSETTES: Place a parchment shape inside a cup section. Place a second cup section on top of parchment. Fit a gold paper lace medallion inside the cup section and fasten all parts together by forcing a corsage pin through the center. See Fig. E. Arrange the petals.

6. ASSEMBLING TREE: Secure rosettes to foam by dipping end of pin into glue and forcing into foam. Start at bottom. Seven rosettes are used for the bottom row. Continue making rows of rosettes up the cone until it is entirely covered. Rosettes may be made smaller for the top by cutting shorter petals. Approximately thirty rosettes are needed to cover the cone tree.

7. Attach a bead spray to top of tree as shown.

Fig. A Cut out single cup.

Fig. B Cut cup into eight parts.

Fig. C Cut ends in to points.

Fig. D Fold and cut parchment.

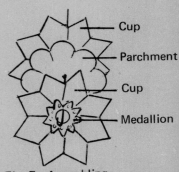

Fig. E Assembling rosette.

Cup
Parchment
Cup
Medallion

Poinsettia

Traditional Poinsettias, painted red or left white, are a gracious accessory to use at Christmas time. This flower is 4" across. See color photo of arrangement on page 37.

MATERIALS

Egg carton, white
8 Poinsettia peps and stamens
26 gauge fine wire, bare
Chenille stem
Acrylic paint, red
Poinsettia leaves
White craft glue
Floral tape, green

DIRECTIONS

1. FLOWER CENTER: Make a cluster of eight poinsettia peps with stamens around them. Allow stamens to extend above the peps approxmiately 1/4".

2. Place a chenille stem along side the flower center and wrap together with fine wire. See Fig. A. Cut off stamen heads which extend below the peps. Wrap stem with tape being sure to pull all stretch out of the tape while using it.

3. PETALS: Cut off lid and extra foam from front of carton. Cut the bottom of carton into three equal sections of four egg cups each. See Fig. B.

4. Cut out a set of four petals, 1/2" wide, from the center of the four cups. See Fig. C.

5. Trim rough edges from four petals and cut them slightly narrower. To make a little longer petal cut a "U" shape in between each petal as shown in Fig. D.

6. Make two sets of large petals and one of smaller petals. To make a smaller petal cut it shorter and narrower. Slit in between all petals to within 1/4" of the bottom of the cup.

7. Petals may be painted red before assemblying the flower.

8. ASSEMBLY: Place small petals on top of two larger ones and glue three layers together. Arrange so all petals show.

9. Push flower center with stem down through the center of the petals. The chenille stem will penetrate the foam easily.

10. Add poinsettia leaves while taping down stem.

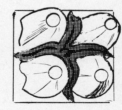

Fig. A Fasten chenille stem to flower center with fine wire.

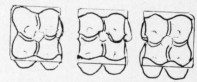

Fig. B Cut egg carton into three equal parts.

Fig. C Cut out cups leaving ridge 1/2" wide for petals.

Fig. D Cut a "U" shape between each petal to elongate it.

Lenny the Lion

The whimsome expression of Lenny the Lion is ready to capture the heart of all who see him. He sits 4" tall and is complete with a wavy tail.

MATERIALS
2 Egg cartons
Clear all-purpose glue-glaze
Acrylic spray paint, golden yellow
Acrylic paint, brown and pink
Glass eyes, green, 7 mm

DIRECTIONS

1. BODY: Cut out the two egg cups at the end of the carton and use hot cutter to fuse the large ends together. Cover remaining holes with other pieces of carton.

2. HEAD: Notice that the eggs are kept apart down the center of the carton by five volcano-shaped dividers. Cut the top 1" off of two of these and fuse them together for the head. Use a curved piece of egg cup to fuse head to body. See Fig. A.

3. SHAPE: Build up the figure into the general shape of a lion by adding additional pieces of carton. Use 1/2" curved sections sliced from the side of the center divider to fill in around the neck. Fuse layers of penny-size pieces cut from the sides of the egg cups to enlarge the figure behind the neck.

4. MANE: Cut the lid into oblong pieces, 1/4"x 1/2" in size. Start fusing these to the head for a mane. Leave the face, below the eyes, bare. The pieces will have to be layered to build up the mane to the proper proportion. Use a curving stroke with the cutter to simulate fur.

Fig. A Fuse the head to body. Notice the ridge for the nose.

5. HIND LEGS: See pattern and cut the hind legs from a triple layer of carton lids. Trim and sculpture toes on the paws. Fuse onto body.

6. FRONT LEGS: Cut these from a double layer of carton lids. Trim and add toes on paws. Melt holes in body and position legs in them. See Fig. B for armature.

7. EARS: Cut these from the corners of the lid. Melt slits where the ears go. Place in position and glue to hold.

8. TAIL: Follow pattern, pg. 22, and cut tail from a double layer of lid. Fuse 1/4"x 1/2" pieces from lid to the end

Fig. B Armature for lion.

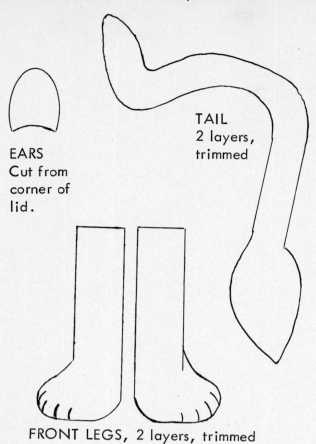

EARS
Cut from corner of lid.

TAIL
2 layers, trimmed

FRONT LEGS, 2 layers, trimmed

of tail. Apply hot cutter to end of tail to simulate hair. Melt a hole in the rump and glue the tail in place. It is glued in sideways so that it will be close to the body.

9. Add clear all-purpose glue-glaze to all stress points --- especially between hind legs and body, to strengthen.

10. Paint with golden yellow acrylic spray paint. Paint face black. Dry-brush mane. Brush tail with brown and add pink to inside of ears.

11. Glue eyes in place. Paint entire lion with clear all-purpose glue-glaze.

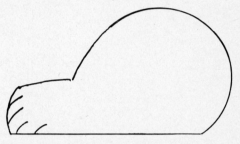

HIND LEGS, 3 layers, trimmed

This stunning wreath was made by covering a white foam wreath with egg carton rosettes. See page 38 for rosette directions.

Benny the Bear

Who wouldn't love this "cutie" bear! He sits 5" high and clutches a colorful flower chain in his paws. His fur is carved with a heated tool.

MATERIALS

3 Egg cartons
Glass eyes, brown, 7 mm
Acrylic spray paint, golden yellow
Acrylic paint, brown and gold
Clear all-purpose glue-glaze

DIRECTIONS

1. BODY: Cut out the two egg cups at the end of the carton and use hot cutter to fuse the large ends together. Cover remaining holes with other pieces of carton.
2. HEAD: Notice that the eggs are kept apart down the center of the carton by five volcano-shaped dividers. Cut the top 1" off of two of these and fuse them together for the head. See Fig. A. Use a curved piece of egg cup to fuse head to body.
3. FACE FEATURES: The carton lid is latched closed by two projections on the bottom half of the egg carton. Fuse one of these projections to the head for a nose and cut in nostrils. Cut a 5/8" round piece from a stack of three lids for the protruding jaw and fuse in place. Cut a hole in the center for mouth. Fuse portions of the volcano-shaped divider to the top of the jaw for a top lip. See sketches below.
4. SHAPE: Fuse additional pieces of carton onto the figure until its shape is like a cute bear. Use 1/2" curves sliced from the center divider to build out the neck. Build out the fat bottom and arms and legs with penny-sized pieces cut from the lid and fused together.
5. FUR: Cut penny-size pieces from the sides of the egg cups. These do not have to be round. Save the curved dividers, between the egg cups and use them to build out around the neck. Start fusing "pennies" to the body, overlapping as needed. Sculpture with hot cutter to simulate fur. Where necessary layer pieces to add dimension such as neck and head.
6. ARMS AND LEGS: These are made by fusing layer upon layers of pennies until they are built out far enough. The arms can be curved by using this same technique.

Fig. A Fuse parts of head together and to body.

Close-up of face features.

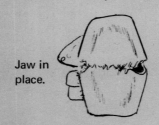

Jaw in place.

Top lip in place.

7. EARS: Cut two ears from the curved corners of the lid. Attach by melting a slit, the same size as the ear, in the head and glue ears in place.

8. Coat all stress points and around body with clear all-purpose glue-glaze. Allow to dry.

9. Spray bear with golden yellow paint. Glue in eyes. Make eye lids the same as upper lips in step 6. Attach eye lids and paint with golden yellow. Paint nose and mouth brown.

10. Completely coat the bear with clear all-purpose glue-glaze for added strength.

11. Any shape small flowers can be cut from colored cartons and strung between bear's paws.

Bear's Ears

Flower Pattern

Hoot Owl Plaque

This wise, old owl is 5" high and perches on a twig to form an attractive 3-D plaque. Your friends won't believe he was once an egg carton!

MATERIALS

Egg carton, any color
Spray paint, beige
Acrylic paint, yellow, white and black
Pair of glass eyes, 12 mm
Twig, 4-1/4" long
Oval wooden plaque, 6-1/2" x 10"
White craft glue

Beak Pattern

Feather Pattern

DIRECTIONS

1. ARMATURE: Cut out two end cups. Fuse together with a hot cutter for owl's body. Cut out ends of two volcano-shaped center dividers and fuse together for head. Fuse head to top of body.
2. Cut body in half lengthwise. See Fig. A. Glue half to a piece of light cardboard. Trim excess cardboard away from the sides with a craft knife. This makes a 3-D type armature with a flat back which can be glued to a wooden plaque.

Feet Pattern

3. FEATHERS: See pattern and cut several triangular pieces from egg carton. Save two curved triangles for owl's "horns". Smaller triangles will be needed for face.
4. FEATHERING BODY: Fuse a row of triangles around bottom of owl. Next row should overlap first row. Alternate feathers. See Fig. B. Continue until body is covered.
5. FACE FEATHERS: These should be applied in two circular patterns, the center of which will be the eyes. See Fig. C. One triangle is placed in the upper center of the face to top off the feathers. Add "horns" to head.
6. BEAK: Use pattern and cut from three layers of lids. Trim, fuse and glue in place.
7. FEET: Use pattern and cut from lid. Fuse and glue into place.
8. PAINTING: Spray owl with beige paint. Dry-brush with yellow, white and black.
9. EYES: Glue glass eyes into place. Melt ends of feathers level so eyes will set flat.
10. WOODEN BACKGROUND: Paint as desired. Glue owl into place sitting on a twig.

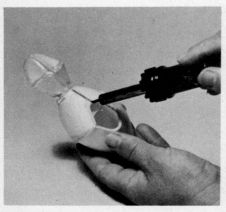

Fig. A Cut body in half lengthwise with a hot cutter.

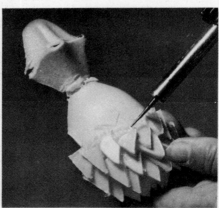

Fig. B Fuse feathers to body.

Fig. C Placement of feathers and features on face.

44

you can make
EGG CARTON FLOWERS

Foreword

You don't have to be an "egg head" to enjoy the exciting new projects so easy to create with pastel colored egg cartons. Each plastic egg carton is guaranteed to be brimming with dozens of great shapes for flowers and many hours of craft fun. Egg carton blossoms have a satin, pearly finish --- gorgeous by itself, or most colorful when painted with acrylics. The soft plastic is easy to cut with scissors or a craft knife. In minutes, ordinary egg cartons can be transformed into daisies, roses, lilies, daffodils, carnations, and many more. Leaf through the following pages and learn the secrets for lovely egg carton bouquets.

These "picture perfect" roses become the pick of the season all year round when they are mounted on a plaque. With such an enchanting way to show off egg carton blooms, you'll be tempted to create a whole gallery full!

Seco Lily

The regal Sego Lily, Utah's state flower, is included by popular demand. This 2½" fragile looking lily is most beautiful made of pearl white egg cartons.

DIRECTIONS

1. Use petal pattern to cut three petals from the corner cups of a white carton.
2. Paint center design on petals with acrylics. Follow color scheme on pattern.
3. Glue center tabs of petals together forming a flower.
4. Cut three calyx and paint them green. Glue to back of flower.
5. Bend small, flat loop on end of 18 ga. stem wire. Insert wire thru flower center. Glue 6 commercial lily stamens around center. Cut tip of lock tab from carton, paint green and glue to center.
6. Cut lily leaves 3/8" wide and add while taping down stem.

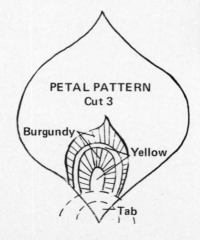

PETAL PATTERN
Cut 3

Burgundy

Yellow

Tab

CALYX PATTERN
Cut calyx from side of egg cup.

Peacock Plaque

The stunning plaque with the colored Peacock, shown on the cover, was made with plastic, turquoise meat trays.   Another variation of the same design is shown on page 6.   The sharp contrast between the white trays and the black background is very dramatic.
The plaque is 18 x 24 inches.

MATERIALS
5 Plastic meat trays, 9"x 11", turquoise or white
3 Plastic egg cartons, turquoise or white
Thick white glue

DIRECTIONS
1. Make a master paper pattern for bird and branches shown below.
2. Pin bird pattern onto a meat tray. Trace with a pencil applying just enough pressure to make visible indentations.
3. Use scissors to carefully cut out the bird. Cutting should be done from all angles as plastic does not cut like paper.
4. Indent details to simulate features.
A potato peeler is an ideal tool for making indentations. See Fig. A.
5. Cut branches from a white tray. Indent bark design.

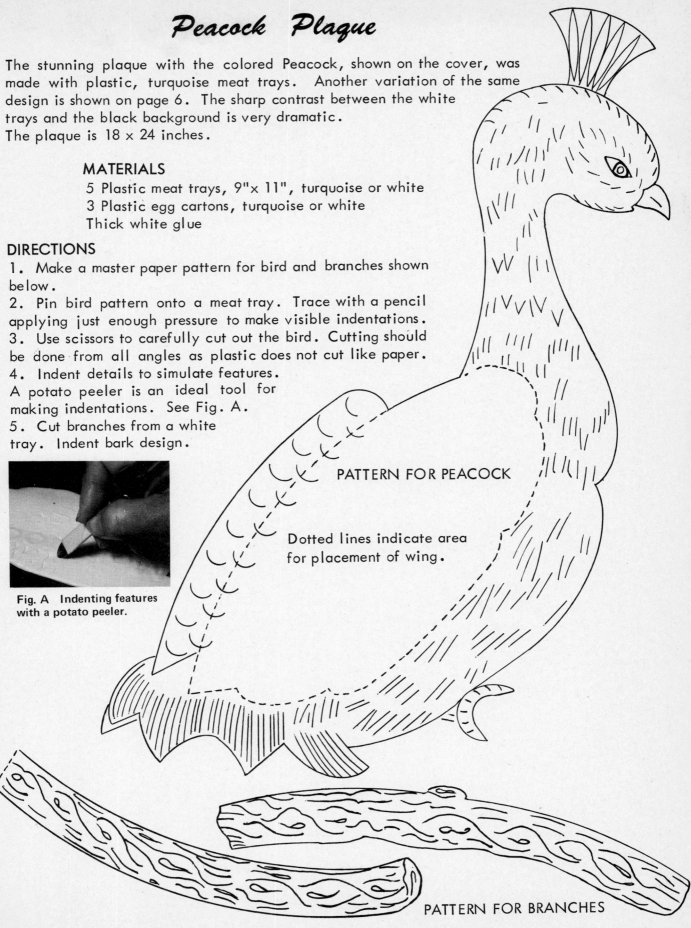

Fig. A   Indenting features with a potato peeler.

PATTERN FOR PEACOCK

Dotted lines indicate area for placement of wing.

PATTERN FOR BRANCHES

6. TAIL FEATHERS AND EYE-SPOTS: Make master paper patterns of all pieces on this page. Cut quantities indicated.

7. Use straight scissors to feather edge of all feathers. Alternate long and short cuts. Always cut toward the top of the feather. See Fig. B.

8. Please turn to page 21 for painting instructions and paint tail feathers and eye-spots.

9. Assemble tail feathers. See Fig. C. Glue a #1 onto each of eight #2 feathers and then glue B eye-spot onto the top of each of these. Glue a single #2 feather onto each #5 feather and then glue a C eye-spot onto the top. Glue a #3 feather onto each #6 as indicated on pattern. Glue a D eye-spot onto the top. Glue a #4 feather onto each #7 and then glue a D eye-spot onto the top. Extra feathers will be used later when assembling bird on plaque.

Fig. C An assembled tail feather.

TAIL FEATHER PATTERNS

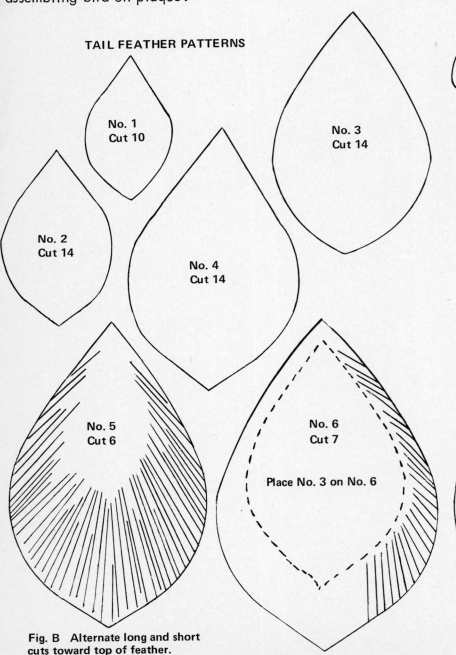

No. 1
Cut 10

No. 3
Cut 14

No. 2
Cut 14

No. 4
Cut 14

No. 5
Cut 6

No. 6
Cut 7

Place No. 3 on No. 6

No. 7
Cut 7

Place No. 4 on No. 7

Fig. B Alternate long and short cuts toward top of feather.

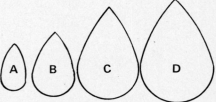

A B C D

EYE-SPOT PATTERNS

Cut from top of volcano-shaped divider between egg cups. See Fig. D on page 5.

Cut 2 A   Cut 13 C
Cut 15 B   Cut 14 D

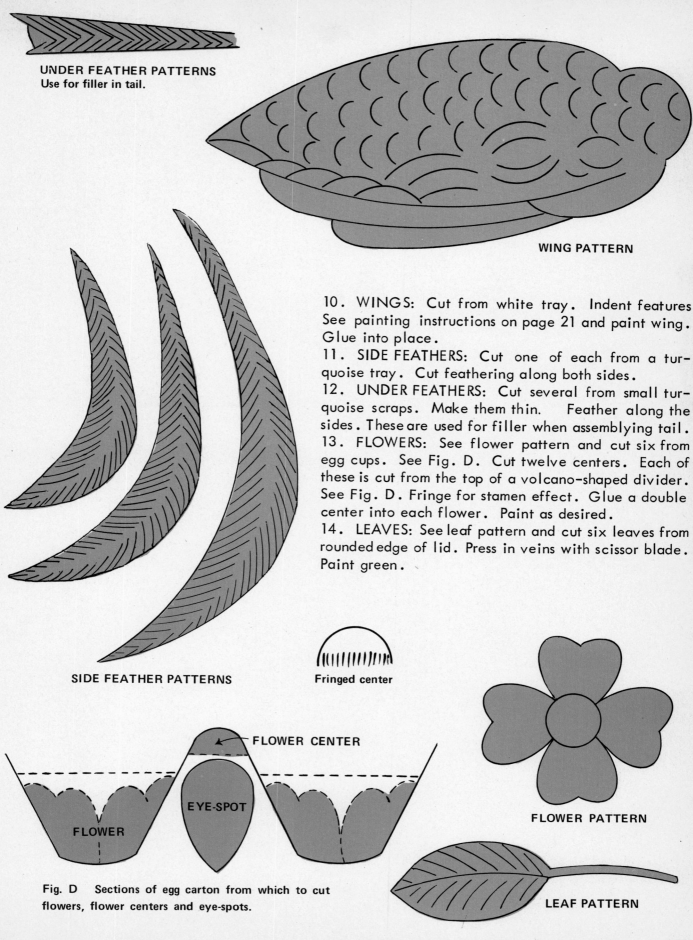

UNDER FEATHER PATTERNS
Use for filler in tail.

WING PATTERN

10. **WINGS:** Cut from white tray. Indent features See painting instructions on page 21 and paint wing. Glue into place.

11. **SIDE FEATHERS:** Cut one of each from a turquoise tray. Cut feathering along both sides.

12. **UNDER FEATHERS:** Cut several from small turquoise scraps. Make them thin. Feather along the sides. These are used for filler when assemblying tail.

13. **FLOWERS:** See flower pattern and cut six from egg cups. See Fig. D. Cut twelve centers. Each of these is cut from the top of a volcano-shaped divider. See Fig. D. Fringe for stamen effect. Glue a double center into each flower. Paint as desired.

14. **LEAVES:** See leaf pattern and cut six leaves from rounded edge of lid. Press in veins with scissor blade. Paint green.

SIDE FEATHER PATTERNS

Fringed center

FLOWER CENTER

FLOWER

EYE-SPOT

FLOWER PATTERN

Fig. D Sections of egg carton from which to cut flowers, flower centers and eye-spots.

LEAF PATTERN

Covering Cardboard Background Plaque For Peacock

MATERIALS

18"x 24" heavy cardboard
20"x 26" velvet, black
White craft glue

DIRECTIONS

1. Be sure 18"x 24" cardboard has straight edges. Heavy cardboard cuts neatly if cut with a craft knife using any good straight edge for a guide.
2. To glue velvet to cardboard, place velvet right side down on a smooth, clean surface. Spray velvet and board with mounting adhesive. Center board on top with an inch of velvet showing on all sides.
3. Run a smooth bead of glue around the edge of the cardboard; also around the edges of the velvet. Fold edges over firmly. Do not turn over or touch the underside while gluing or finger marks will show.
4. See Fig. A and cut away excess from corners.

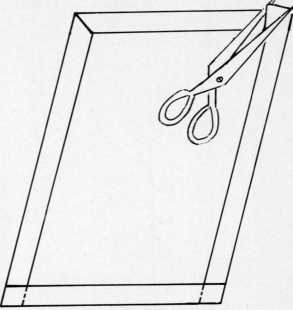

Fig. A Cut excess away from corners.

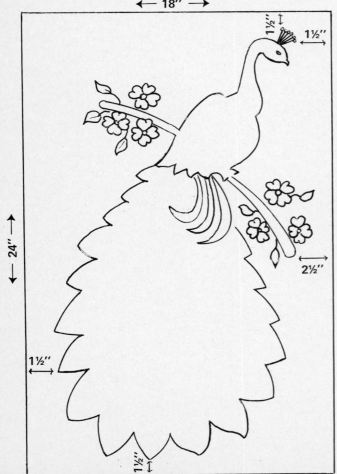

Mounting Peacock On Plaque

See guide, at left, and assemble peacock parts onto velvet-covered cardboard background. When effect is as desired, carefully secure each piece into place with white craft glue. Use filler feathers, as needed on tail.

Instructions For Painting Peacock Plaque

The elegant Peacock Plaque can be made in color, as shown on the front cover --- or all white and mounted on a dark colored background as shown on page 52. Construction instructions for the plaque begin on page 47.

For color, cut Peacock from turquoise meat trays and paint before assemblying. Stencils are used to facilitate painting portions blue and green.

The spray colors used for painting are:

Transparent Blue   Opaque Grey
Transparent Green   Wood Tone
Royal Blue   or Gold

FIG. C   STENCIL FOR PEACOCK'S EYE

SPECFIC INSTRUCTIONS FOR PAINTING

FEATHERS ON BODY: Spray feathers with both transparent blue and transparent green. Spray a little more green in some places and mostly blue around the neck near the head.

TAIL FEATHERS: Spray with transparent blue and green. Wood tone color is used for gold effect. To apply, cut stencils as shown in Fig. A. Place on feathers of corresponding number and spray with wood tone. Remove stencil. Hold spray can farther away and lightly spray fringed feathers with a fine mist of wood tone.

TAIL FEATHER EYE-SPOTS: Spray paint with transparent blue. Cut eye-spot stencils as shown in Fig. B. Place into position on eye-spot and spray royal blue.

BEAK AND AROUND EYE: See Fig. C and cut stencil. Paint eye with wood tone color. Add dark dot when dry. Paint beak with wood tone color.

FIG. A   STENCILS FOR TAIL FEATHERS

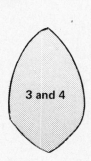

1   2   3 and 4   5   6 and 7

FIG. B   STENCILS FOR TAIL FEATHER EYE-SPOTS

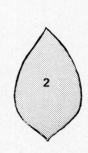

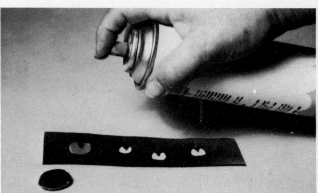

Use stencil for spraying tail feather eye-spots.

Here is an elegant bird perching on a slender piece of "bark". Shades of blue and green stand out dramatically from the black background. An 18"x24" plaque has been covered with black velvet --- providing the luxurious contrast this splendid peacock deserves. The peacock plaque is certain to grace any living room with a tasteful accent.

PEACOCKS CAN BE MADE IN COLORFUL TURQUOISE OR STRIKING EGG CARTON WHITE

Look how white egg cartons and meat trays have been combined with black to create a reverse silhouette of the proud peacock. The pearly white bird is shown off on an 18"x 24" plaque which has been covered with black velvet. Perky detailing has been added with the clever fringing and indentations on the feathers.

"A rose is a rose ..." but this egg carton beauty is something special. The large yellow rose measures 5" in diameter and is surrounded by several budding young beauties. A generous assortment of egg carton greenery is used as filler around the flowers. Note the goblet-like vase used to display this spring-time sampling of egg carton "garden" finery.

Assorted dahlias make an impressive contribution to interior decor. One large dahlia, 6½" in diameter, is accompanied by a cluster of buds and foliage. A slender Oriental vase displays this bouquet perfectly. Consider a bright splash of pink dahlias for a bare piano top.

Full-Bloom Rose

Egg cartons in any of the delicate pastel colors will suit this rose just fine. To make this rose "blush" happy, just surround it with a bevy of small buds.

MATERIALS
8 Egg cartons, color of your choice
18 ga. stem wire, covered
30 ga. wire, uncovered
Floral tape, green
Small pins, 3/4"
White craft glue
All-purpose spray paint,
    avocado or moss green

DIRECTIONS

1. ROSE BUD: Cut a 2/3 round section from a corner cup as shown in Fig. A.   Soften top edge by pressing and squeezing with fingers. Glue bottom tips and roll tightly. Pin at top to hold. Insert an 18 ga. stem wire and wrap bottom tightly with 30 ga. wire. See Fig. B.

2. Cut three small petals. Cut 1/4 round section from corner cups. See next page for pattern. Add glue to tips and sides as indicated in Fig. C. Add petals, one at a time, around center. Pin near bottom to hold petals closed. To complete a bud add a calyx. See Fig. D on next page for instructions.

MEDIUM-SIZE ROSE:
1. Make a bud as explained above and use for center of rose.
2. Cut five medium size petals. Cut 1/2 to 1/3 round section from corner cups. See pattern on next page.
3. Add glue to tip and sides. Add, one at a time, for third row of petals. Spiral by placing one edge over and one edge under the petal next to it. Pin at bottom to hold petals closed. Add a calyx.

FOR FULL-BLOOM ROSE:
1. Make a medium-size rose.
2. Cut twelve large petals. Cut 1/3 round section from corner cups.   See next page for pattern.   Add glue to tip and sides. Make a rose of six petals adding one at a time. Pin petals at bottom. Repeat for another row of six petals.
5. CALYX: Cut six sections from the center divider. See Fig. D. Paint green. Glue and pin calyx to back side of rose.
6. LEAVES: Make basic leaf, as shown on next page and tape to rose stem.

Fig. A   Cut 2/3 cup so bottom is pointed.

Fig. B   Secure stem to bottom of rose center with 30 ga. wire.

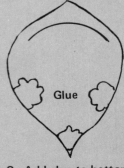

Glue

Fig. C   Add glue to bottom tip and both sides of petal.

ROSE PETAL PATTERNS

The rose petals are cut from the end egg cups only. The relative size of each petal is shown below.

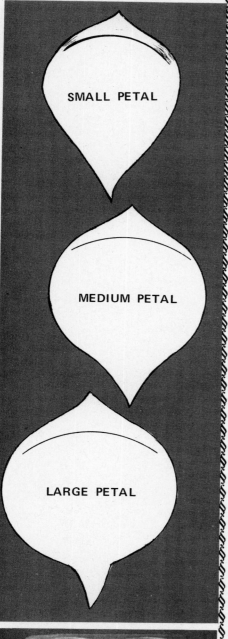

SMALL PETAL

MEDIUM PETAL

LARGE PETAL

Fig. D CALYX

The calyx is cut from the center divider. Cut six for each rose. Paint them green.

Fig. B Assembled basic rose leaf.

Basic Rose Leaf

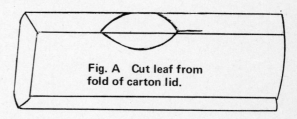

Fig. A Cut leaf from fold of carton lid.

DIRECTIONS

1. Trace paper patterns and cut leaf sections out of the fold on the carton lid. See Fig. A. Use different sizes for various size leaves.

2. Mark in veins with a scissor blade.

3. Insert darning needle to make a hole for easy installation of wire.

4. Insert and glue #22 ga. wire 1" into leaf for reinforcement. Allow 1" to extend for joining three sections together.

LEAF PATTERNS

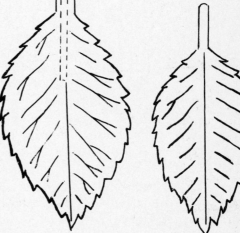

5. Assemble and tape the three leaf sections together to form a basic rose leaf. See Fig. B.

6. Paint leaf green.

Dahlia

Although dahlias are native to cool climates, egg carton artistry lets you capture all of their colorful beauty, no matter where you live. Sturdy, tall wire stems are used to fasten these generous-sized dahlias. They measure 6½" in diameter. Larger dahlias join well with buds, and both have a particularily pearly-like shimmer when made of pink egg cartons. See color photo on page 53.

MATERIALS

6 Egg cartons, color of your choice
18 ga. wire, covered
10 pcs. 22 ga. wire, covered
3 pkg. 24 ga. wire, covered
White foam ball, 1/2" dia.
All-purpose acrylic spray, avocado or moss green
Floral tape, green
White craft glue

DIRECTIONS

1. Cut and trim foam ball as shown in Fig. A. Make a small loop in the end of the 18 ga. wire. Push wire through the center of the ball and bend the loop against the top of the ball.

2. PETALS: Make master paper patterns. Cut petals from sides of egg cups. Need 60 A petals, 30 B petals, 30 C petals and 20 D petals.

3. Fold bottom of petal together and wire the bottom tip with a 3" length of 24 ga. wire. See Fig. B. Wire all A, B and C petals.

4. ASSEMBLY: Dip end of petal wires in white glue and insert into foam ball. Place about fifteen of the largest petals around the trimmed part of the ball as in Fig. C. Place 1/4" apart. Repeat for two more rows. Then decreasing in number and size, fill ball until 1" of ball remains exposed at the top.

PETAL PATTERNS

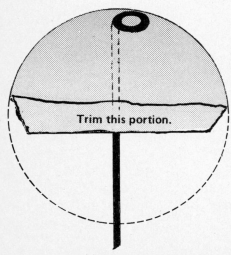

Fig. A Cut off the bottom 1/3 of a foam ball and taper sides. Push stem wire through the center so loop rests on top of the ball.

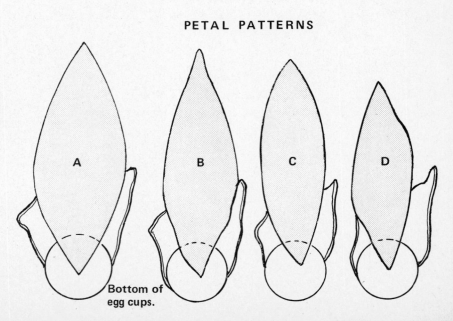

A   B   C   D

Bottom of egg cups.

56

5. CENTER: Wire 10 D petals together as shown in Fig. D. Insert into top of ball. Make certain that the center petals fit evenly with the other petals. If needed, glue and pin more D petals into place.

6. CALYX: Make master paper pattern. Cut two; each from the center of a carton lid. Glue both pieces together so petals alternate. Spray green and dry. Make a hole in the center and draw through stem to back of Dahlia. Secure with tape.

7. LEAF: Make master paper pattern for leaf. See Fig. E for section from which to cut leaves. Make a hole by running a darning needle up stem into leaf. Glue a 5" piece of 22 ga. wire into this hole for leaf stem. Cut two for each flower. Spray green and tape to flower stem.

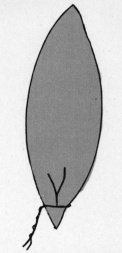

Fig. B Wire bottom of petal.

CALYX PATTERN

Fig. C Top view of fifteen petals inserted in ball for first row.

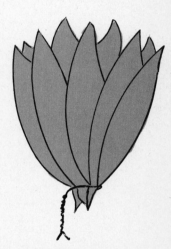

Fig. D For Dahlia center, wire ten D petals together. Trim off bottoms.

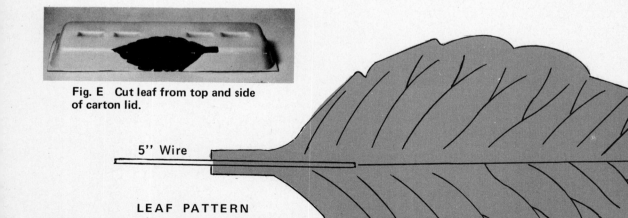

Fig. E Cut leaf from top and side of carton lid.

5" Wire

LEAF PATTERN

Make a hole in the stem end by inserting a darning needle. Glue a 5" wire into hole for a stem.

Rose Plaque

White, yellow and pink roses show a pretty variation of a velvet covered plaque. The delicate pastel colors of the roses show up beautifully against the rich, textured dark background. Gold cording adds an extra highlight to the framework. The open roses measure 2½" in diameter and a variety of bud-sized roses add interest. Rose petals may be shaded with paint for more colorful detailing. A pair of egg carton flower plaques make a unique wall display guaranteed to reap compliments from season to season.

TRANSVAAL DAISIES, in delicate pastels, are easily made following instructions on page 66.

Carnation Plaque

A brown velvet covered plaque, 13½"x17½", provides a handsome background for pretty egg carton carnations. Large white and pink carnations measure 2½" in diameter and are accompanied by several buds, peeking from the background. Pinking shears have been used for serrating the edges of the petals, giving flowers a life-like appearance. Gold spray paint and a scalloped border nicely transform an ordinary plastic bottle into an antique looking vase: just perfect for holding these fancy carnations.

Carnation

DIRECTIONS

1. Cut master paper patterns of petals. Lay pattern on side of an egg cup and cut out three # A petals. Notice bottom tip is bent at right angle. Serrate top of the petals with pinking shears.
2. Put glue on bottom tabs of petals Roll one into a tight roll. Spiral other two petals tightly around it. Wrap bottom with 30 ga. wire. See Fig. A.
3. Cut five more # A petals. Spiral and add as a second row around center. Glue and wire to hold.
4. Glue end of a 12" stem wire into bottom of flower. This step is not needed for carnations used on a plaque.

MATERIALS

Egg carton, color of your choice
White craft glue
30 ga. wire, covered
18 ga. stem wire, covered
Floral tape, green
All-purpose spray paint, green
Pins, 3/4"

PETAL PATTERNS

PETAL
A

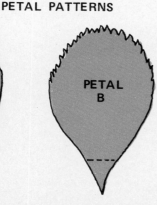

PETAL
B

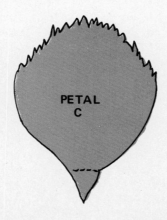

PETAL
C

Fig. A   Bottom of carnation center secured with fine wire.

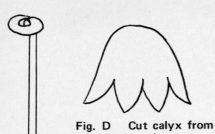

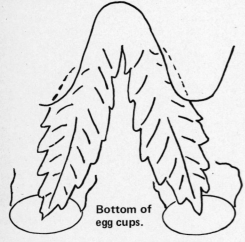

Fig. C   Bend a flat loop in end of stem wire.

Fig. D   Cut calyx from volcano-shaped center divider.

Fig. E   Cut a four-section leaf from volcano-shaped center divider.

Bottom of egg cups.

5. Make a small, flat loop in one end of the stem wire. See Fig. C. Punch a small hole in center of daisy. Insert wire and bring the looped end down into the cup. Glue to hold.

6. Cut a round "button" from top of volcano-shaped center divider. Lightly spray brown. Cover loop by gluing "button" to center of daisy.

7. CALYX: Fig. D. Cut calyx from volcano-shaped center divider. Spray green and allow to dry. Make a hole in the center and push up stem to back of flower. Tape stem. Be sure to pull all stretch out of tape while using it.

8. LEAVES: Fig. E. Cut a four-leaf cluster from volcano-shaped center divider. Serrate edges. Vein both sides with an open blade of scissors. Spray leaves green. When dry, make a hole in the center of the cluster and push up stem wire. Tape to hold.

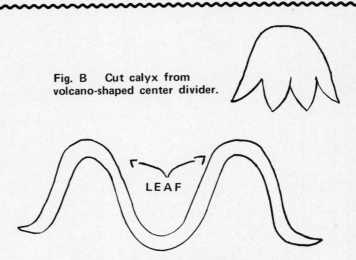

Fig. B   Cut calyx from volcano-shaped center divider.

LEAF

Fig. C   Cut leaves from volcano-shaped center divider.

CARNATION --- continued from page 59

5. Cut five # B petals. Spiral and glue them around last row. Pin at bottom to hold.

6. Cut seven more # B petals. Glue and pin them around flower.

7. Cut seven # C petals. Glue and pin them around flower.

8. Another row of nine petals can be added for a fuller flower.

9. CALYX: Cut from volcano-shaped center divider. See Fig. B. If center divider is too wide, cut a slit in calyx and lap it over. Glue and pin until dry. Add to back side of flower. Tape to hold in place. This step is not needed if carnation is used on plaque.

10. Cut two leaves from volcano-shaped center divider. See Fig. C. Spray green and allow to dry. Punch a hole in center of leaf and insert stem wire. Tape stem and add other leaf. See Fig. D.

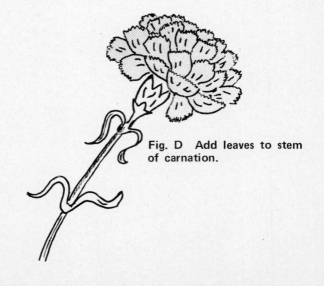

Fig. D   Add leaves to stem of carnation.

Forget-Me-Nots

Light blue egg cartons are cleverly snipped to make small bunches of forget-me-nots. Each flower is ¾" in diameter. There are seven per cluster.

DIRECTIONS

1. For each blossom cut a 3/4" circle from the top of the volcano-shaped center divider of a blue egg carton.

2. Divide circle into five equal parts by cutting almost to the center. Fig. A. Form petals by rounding divided sections. Fig. B. To cup blossom hold cut-out in palm and press the head of a corsage pin against the center.

3. Cut out a tiny yellow center. (A small paper punch is excellent.) Glue to center of blossom.

4. Cut four 5" lengths of 28 ga. wire. Bend in the middle. Make a small loop on each end. See Fig. C. Wrap these four wires in a cluster to a 4" piece of 18 ga. stem wire. Wrap 1" down stem with floral tape.

5. Glue a blossom onto each wire loop.

6. Cut a four-leaf cluster from the volcano-shaped center divider of a yellow carton. See Fig. D. Vein both sides of leaves with an open scissor blade. Spray leaf cluster green.

7. Punch a hole in the center of the leaf cluster and push up stem wire. Tape to secure.

MATERIALS

3 Egg cartons, blue
1 Egg carton, yellow
28 ga. wire, covered green
Stem wire, 18 ga.
Floral tape, green
White craft glue
Spray paint, avocado or moss green

Fig. A   Divide 3/4" circle into five equal parts.

Fig. B   Round off petals.

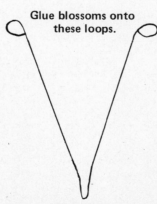

Glue blossoms onto these loops.

Fig. C   Make small loop on end of each wire.

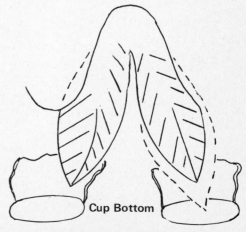

Cup Bottom

Fig. D   Cut four-cluster leaf.

With a little imagination, a dramatic floral performance can be staged in front of this egg carton screen. A sunny arrangement of daisies becomes the star of the "show". The black egg carton screen has been trimmed with fancy round and star shaped sequins. An antique gold wooden border frames the screen. Hinges join the three partitions allowing for variations in positioning. The screen measures 24" wide and 12" high. White or yellow daisies, arranged in a favorite bowl such as the gold one shown, can be displayed as proudly as any art masterpiece in front of this novel background. What an elegant way to let egg carton bouquets make their debut!

There's no more charming way to liven a tabletop, than with a nosegay centerpiece. This one features spring's loveliest pink roses, forget-me-nots and a bevy of white daisies. A happy combination of pastel colors! A sampling of various shaped leaves serve as filler and add a dash of greenery. This cheerful nosegay is sure to brighten that lone coffee table or stereo, too.

Daffodils

Egg carton daffodils are just in time for fancy spring bouquets! Add a bit of greenery and these long-stemmed yellow blooms make a pleasing arrangement by themselves or combine daffodils with other egg carton florals to create a multi-colored assortment of choice springtime blossoms. Yellow daffodils have been shaded with orange and brown. Petals are bent to various angles.

MATERIALS

Egg cartons, yellow
White craft glue
Powdered concentrated dye, orange
Rubbing alcohol
All-purpose spray paint, avocado green and light brown
Stem wire, 18 ga., covered
22 ga. wire, covered
Floral tape, green

DIRECTIONS

1. Cut an egg cup from the carton. Cut four petals as shown in Fig. A. Need six petals.

2. Cut the round piece from the bottom of an egg cup. Glue six petals around the edge of it. See Fig. B.

3. Cut the trumpet from the volcano-shaped center divider. Cut well into the adjoining divider for fullness. Ripple the edges. See Fig. C.

Fig. A Cut four petals from each cup.

Fig. B Glue six petals around circle cut from bottom of cup.

Fig. C Cut daffodil trumpet from volcano-shaped center divider.

4. Dissolve a little orange concentrated dye in a small amount of rubbing alcohol. Paint daffodil trumpet with this solution for a transparent, natural shade of color. For solid color use acrylic.

5. Glue trumpet to center of petals. Allow glue to set.

6. Make a small, flat loop at one end of piece of stem wire. See Fig. D. Pull wire through center of daffodil trumpet and circle of petals. Glue loop flat inside daffodil trumpet.

7. CALYX: Cut from the bottom half of cup. See Fig. E. Spray green. Allow to dry. Push up stem wire to back of flower. Tape to hold.

8. STAMENS: Cut stamens from the volcano-shaped center divider. See Fig. F. Bend tips back. Spray paint tips only a light brown by wrapping bottom portion in a paper napkin with tips extending. Use 1/2 portion of stamens and glue into place so wire loop, inside trumpet, is covered.

Fig. D   Shape a small, flat hook on end of stem wire.

9. LEAVES: Make paper pattern. Cut each leaf from the fold of a carton lid. See Fig. G. Spray green and allow to dry. Make a hole in stem end by inserting a 22 ga. wire into the hole for a stem. Add leaves to daffodil stem while wrapping with tape. Be sure to pull all stretch out of tape when using it.

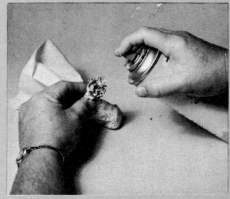

Spray tips of stamens brown.

Fig. E   Cut calyx from bottom of egg cup.

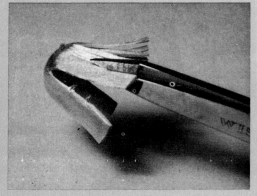

Fig. F   Cut long stamens from volcano-shaped center divider.

Fig. G   Cut leaf from fold of carton lid.

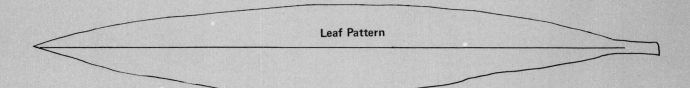

Leaf Pattern

Daisy

Daisies do tell that springtime is flower time. Yellow or white egg carton daisies are a sunny promise for any decor. Petals curve upward around a yellow "button" center and a cup of four leaves surrounds each flower. The daisy is 2" in diameter. Leaf edges are serrated and sprayed with moss green colored paint.

DIRECTIONS

1. Cut out an egg cup and trim off enough of the top to make it even all around.

2. Cut a cup in four equal parts and then further divide each into thirds. See Fig. A. You will have twelve petals.

3. Make a flat hook at one end of a 6" length of chenille stem. See Fig. B. Push stem down through center of daisy until the hook rests on the blossom. Glue hook to hold blossom.

4. Cut a round "button" from top of volcano-shaped center divider. Lightly spray brown. Glue into center of daisy.

5. Cut a calyx from bottom of egg cup. See Fig. C. Spray green. Make hole in center and push up stem to back of flower.

6. Tape stem. Be sure to pull all stretch out of tape while using it.

7. Cut a four-leaf cluster from the volcano-shaped center divider. Serrate edges. See Fig. D. Vein both sides with an open scissor blade.

8. Spray leaves green. When dry, make a hole in the center of the leaf cluster and push up stem wire. Tape to secure.

MATERIALS

Egg carton, white or yellow
Chenille stem
White craft glue
All-purpose spray paint, lt. brown
    and avocado or moss green
Floral tape, green

Fig. A   Cut egg cup into twelve petals.

Fig. B   Make a flat loop on end of chenille stem and bend at right angle.

Fig. C   Cut calyx from bottom of egg cup.

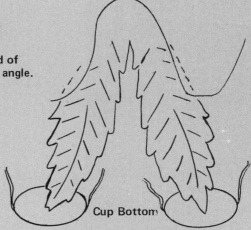

Cup Bottom

Fig. D   Cut four-cluster leaf from volcano-shaped center divider. Serrate edges.

Transvaal Daisy

Transvaal daisies are bursting with yellow, lilac, and pink colors. Mingling by the dozens in a large ceramic bowl, they make a happy mixture of pastels. This special daisy measures 4″ in diameter. The spindly, slender petals and long stems create an airy effect in the arrangement shown in color on page 58. Brown centers add a splash of contrast to each fluffy flower.

MATERIAL FOR THREE DAISIES:
6 Egg cartons, color of your choice
3 pieces 18 ga. stem wire, covered
Spray paint, all-purpose: lt. brown
    and avocado or moss green
Floral tape, green
White craft glue

DIRECTIONS

1. PETALS: Cut carton into three four-cup sections. From each piece cut an eight-fingered section. See Fig. A. The four fingers that extend into the bottom of the cups are slit into four petals. The four fingers that are cut from the divider are slit into two petals. If you cut the fingers as wide as possible your petals should now appear as shown in Fig. B.
2. Cut ends of petals into points.
3. Repeat for a second section of petals.
4. Glue the two sections together to make a full daisy. Turn the top section so the curved petals are over the straight petals.

continued on page 60

Fig. A   Cut petal sections out of volcano-shaped center divider.

Fig. B   Cut curved sections into two petals. Cut other sections into four petals.

FOLDED MAGAZINE NOVELTIES

WHICH MAGAZINES FOLD THE BEST

We chose the Reader's Digest for most of our figures because its 5½" x 7½" page size and its generous 250 pages-plus make a good full-bodied folded figure. However, we used larger size magazines, such as the Redbook, for the Three Kings. You may want to glue two or more folded magazines together to achieve the proper thickness. Catalogs, phone directories and even pocket-size books have good possibilities, so do not hesitate to experiment with them. We used an assortment of different size magazines for the Contemporary Bell Hanging seen on page 16.

A BIT ABOUT FOLDING

The different folds used are quite simple to do. Folding the pages is a lot of fun and a great craft for T.V. watching, Instructions for folding are included with each project. To make uniform folds, fold over a template. To make one, follow the scale diagram included in the instructions and cut a lightweight cardboard pattern the size and shape of the finished folded page. Simply place this template on top of the first fold and fold over it.

SELECTING THE HEAD

Distinct personality is given to individual projects by topping them off with attractive doll heads. When you choose a head, be sure that the size is in proportion to the body. There are three kinds of heads available:

1. <u>Commercial Doll Heads With a Wire Stem</u> are easy to use. Simply glue the stem into the hole found at the top of the magazine when the magazine is formed into a round body.

2. <u>Commercial Heads Without a Stem</u> often have an opening at the bottom. You can add a stem by stuffing a white foam block into the head. Insert and glue a doubled piece of chenille stem into the foam, allowing 2" to 4" to extend at the bottom. Glue the wire into the hole at the top of the magazine body.

3. *Foam Ball Heads.* To attach the head push and glue a 4" to 6" length of doubled stem 2" into the foam head and then glue the extending stem into the hole formed at the top of the magazine body — OR — press a ¼" dowel into the foam ball and glue the dowel along the binding of the magazine. □

Black Anise

Don't let the jet black color or be-witching green eyes scare you --- Black Anise is really a gentle sort of cat. And, you're in the luck to have found her! She's so quick and easy to make for Halloween parties or as a "purr-fect" treasure for cat fanciers. With extra fluffy chenille paws and ears, natural-like plastic whiskers and a bright orange bow tie, you must admit, Black Anise is really the cat's meow!

MATERIALS

Reader's Digest magazine
4'' dia. White foam ball
4 Chenille bumps, black
2 Chenille bumps, pink
2 Chenille stems, 1 red & 1 pink
Large daisy pep, black
Pr. 12 mm glass or moving eyes
2''x 2'' Felt, lt. green
12'' Curly chenille, black
2 Jumbo chenille stems, black
2/3 yd orange ribbon, 5/8'' wide
6 Bristles from a brush
Craft spray paint, black
White craft glue

DIRECTIONS

1. Remove the front and back covers from a Reader's Digest.
2. Fold each page, for the body, the same as for Pooh, the Pooch, as shown on page 75.
3. Spray the foam ball head and folded magazine body black. Allow the paint to dry.
4. Attach the head to the body with a 4'' length of black, jumbo chenille. Push and glue 2'' into the foam head and then glue the extending 2'' into the hole at the top of the magazine body.

Follow this picture for trimming Black Anise.

TRIMMING BLACK ANISE

EARS: Black and pink bump chenille
HEAD: 4'' dia. foam ball, black
EYES: Green felt and glass eyes
NOSE: Black daisy pep
WHISKERS: Plastic bristles
MOUTH: Pink and red chenille stems, see Fig. A
BOW: Orange ribbon
BODY: Folded magazine, black
TAIL: 12'' Curly chenille, black
BACK FOOT: 6'' Jumbo chenille, blk.
FRONT FOOT: 4'' Jumbo chenille, blk.

Fig. A Close-up of mouth made of pink chenille stem with red chenille tongue. Actual size shown.

EYE PATTERN
Cut 2 of green felt

YOU CAN MAKE
THREE CUTE DOLLS
USING THE SAME
FOLD!

* **GRADUATE**

* **ANGEL**

* **OR BRIDE**

Sweet Girl Graduate

As the last notes of "pomp and circumstance" fade and you're a bit misty-eyed with nostalgia --- it's time to celebrate and shower your graduate with pride and accolades. The Sweet Girl Graduate doll makes an attractive centerpiece for the table; especially when her robe features your graduate's school colors. Later, present her to your guest-of-honor as a lovely keepsake of the occasion.

MATERIALS

Reader's Digest magazine
Doll head, 2¾", stemmed
2 shts. Construction paper, black
1 sht. Construction paper, white
Jumbo chenille stem, white
Small tassel of string
½ yd. Black ribbon, ¼" wide
White craft glue
Spray paint, black or color of
    your choice

DIRECTIONS

1. **GOWN:** Remove the front and back covers from a Reader's Digest. Fold each page into a three fold, as shown in the illustration below. Crease each fold completely. A diagram for a template is given if you want to make the second and third fold around it.

2. Glue the front and back pages of the folded magazine together. This will form a circular gown as shown in Fig. A. Spray paint the gown black or a school color of your choice. Allow the paint to dry.

3. **SLEEVES:** See Fig. B and cut sleeves from black paper. Overlap and glue A edges together forming a cone-shape sleeve. Glue point B, of each sleeve into place on the gown.

4. **ARMS:** Bend a 6'' length of chenille stem into an arm shape as shown in Fig. C. Glue the end of an arm into each sleeve.

5. **COLLAR:** Cut a 3½'' dia. circle of white paper. The collar is held in place by pushing the wire stem attached to the head, thru the center of the collar and down into the hole at the top of the gown.

6. **CAP:** Cut a ½''x 6'' black paper band and glue it into a circle. Glue a 3''x 3'' square of black paper onto the top of it. This will form a cap. Glue the cap onto the top of the doll head. Add a tassel of string.

7. **DIPLOMA:** Cut a 2¾''x 2'' piece of white paper and roll it lengthwise. Tie it with a narrow black ribbon and glue it into a chenille hand.

8. Glue a tiny black ribbon bow to the front of the collar.

Fig. A   Magazine folded into a circular gown.

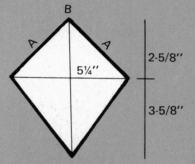

Fig. B   Cut two sleeves of black paper.

Fig. C   Bend chenille into arm shape.

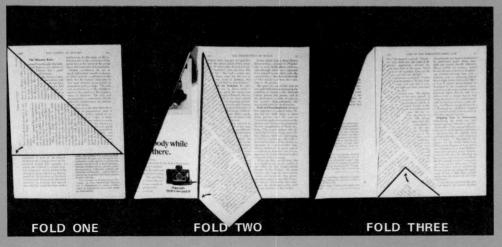

FOLD ONE   **FOLD TWO**   **FOLD THREE**

To form the gown for the graduate, fold each page of the magazine in the sequence shown above. This fundamental three-fold is also used for the bride and angel shown on page 72.

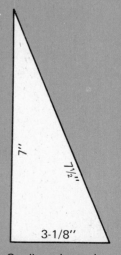

Cardboard template

The same basic fold and instructions are used for the ANGEL and BRIDE as were used for the GRADUATE on pages 70 and 71.

ANGEL

1. Make the sleeves of white paper in the same manner as for the graduate.
2. Spray the angel's gown and sleeves gold.
3. Glue gold paper lace trim down the front of the gown, around the sleeves and across the hair.
4. Place a 3½'' diameter gold paper lace medallion around the neck for a collar.
5. Pin embossed gold paper wings to the back.
6. Add a chenille halo for finishing touch.

BRIDE

1. Spray the magazine gown with white paint.
2. Gather a 9''x 30'' length of white nylon net into an over-gown and secure it around the neck.
3. Glue chenille arms with net sleeves into place.
4. Add a lace collar around the neck.
5. Fold a 6''x 42'' length of net in half for a train and pin it to the top of the head. Use lace and fancy pins to make the headpiece.
6. Add eighteen pearl corsage pins down the front of the gown.
7. Trim the skirt with medallions cut from paper doilies and with strings of pearls.
8. Complete the bride by fastening a nosegay of tiny flowers in her hands.

Birthday Fun

PARTY IDEA

Set Pooh, the Pooch on top of a box full of party favors with a red ribbon attached to each favor. Extend a ribbon to every place setting so each guest can pull out his own favor when the box lid is removed. Of course, you will want to make cute Candy Cup Pups to fill with goodies to complete the ensemble. See page 74 and 75 for instructions.

Flat-Foot, the Clown is always a fun attraction for any child's party. Make him as gay and bright as you wish. The basic instructions are found on pages 76 and 77.

Pooh...
The Pooch

Pooh, the pooch is a happy people lover, especially when he is included in party plans for that special little guy or gal. He'll amuse all the guests with his big, black floppy felt ears, sleek black tail, and mischievous moving eyes. Place him on a decorated box to use as a centerpiece for your next party table. Favors can be fastened to ends of ribbons in the box and run to the candy cup at each place. See

MATERIALS
4''x 6¼'' White foam egg
Reader's Digest magazine
Two 20 mm moving eyes
1''x 1'' felt, red
9''x 12'' felt, black
2 Daisy peps, black
4 Large glass pins, black
2 Chenille stems, black & white
White craft glue
White spray paint
15'' of ½'' Ribbon, red

DIRECTIONS

1. BODY: Remove the front and back covers from a Reader's Digest. See the illustration at the bottom of the next page and fold each page into a three-fold. Use a template, if desired.

2. Lay the pooch body flat and spray well with white paint. Allow to dry.

3. HEAD: To shape the head, cut a V-shape out of the egg head near the pointed end. See Fig. A. Cut a thin slice off the large end, at the back of the head so the egg will stand upright. Rub a piece of white foam scrap against the head to sand and shape it.

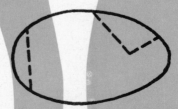

Fig. A Side view of foam egg head showing cuts.

4. See the overlay patterns and cut black felt legs, ears, tail and a red felt tongue.

5. Pin the tongue into place. Glue two black wool daisy peps above the tongue for a nose.

Candy Cup Pup

It's puppy love at first sight when children discover these cute little candy cup pups at your child's party! This little pup reminds them of one of their favorite cartoon characters. His little tummy is a white egg carton section that you fill with candies or nut treats. The oversized black ears, tail, and big feet are black felt. Colorful glass pinheads are used to bring a whimsical expression to the white foam head. Kiddies will "wow" this take-home party favor!

MATERIALS

1½"x 2-1/8" white foam egg
Plastic egg carton cup, white
4 Large head pins, black
1 Large head pin, red
2" Chenille stem, white
4"x 4" Felt, black
12" of ¼" Ribbon, red

DIRECTIONS

1. Cut a thin slice off the large end of the foam egg head so it will set upright.
2. Glue 2" chenille stem into egg. Allow 1½" to extend as shown.
3. See patterns and cut black felt ears, legs and tail. Pin the ears to the side of the head. Add black pin eyes and a red pin nose.
4. Glue the legs and tail to the extended end of chenille. Glue an egg carton cup on top of them.
5. Pin a small ribbon bow under the chin. ☐

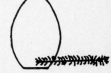

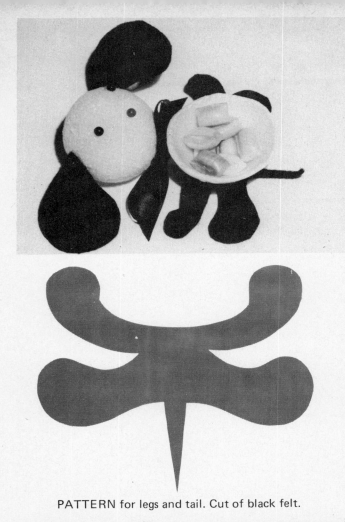

PATTERN for legs and tail. Cut of black felt.

POOH, THE POOCH --- Continued

Glue moving eyes into place. Then glue a 2" curved black chenille eyebrow above each eye.

6. Pin an ear to each side of the head with black headed pins.

7. ATTACH THE HEAD TO THE BODY: Fold the white chenille stem into thirds and push one end 2" into the foam head, in the same manner as for the Candy Cup Pup. Allow 2" to extend. Glue the 2" stem to the binding.

8. Glue the legs into place between the folds of the magazine. Glue the tail to the underside at the end of the magazine body.

9. Pin a red ribbon bow to the neck.

10. Glue pooch on the lid of a goodie box. ☐

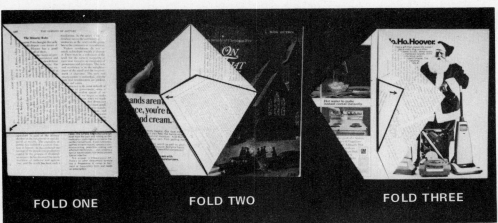

FOLD ONE   FOLD TWO   FOLD THREE

To form the body for the pooch, fold each page of the magazine in the sequence shown above.

Diagram for cardboard folding template.

4½"
3¼"
4¾"
1½"

Flat-Foot, the Clown

Birthdays become a circus of fun when Flat-Foot the Clown joins the merriment! This jolly character is "clad" in red and blue and sports a black top hat for fun. For an added good measure he brings a happy colored balloon along!

MATERIALS
Reader's Digest
Clown head, 2½'' dia.
6''x 8'' Thin cardboard
2 Chenille stems, blue
2 Chenille stems, white
6 Bump chenille, red
1½'' dia. White foam
    ball
6 yds. of 5'' nylon
    net, red
Spray paint, blue
18 ga. wire, covered
White craft glue
Rubber band

DIRECTIONS

1. BODY: Remove the front and back covers from a Reader's Digest magazine. Fold each page as for Pooh, the pooch; see the preceding page. Glue the back and front of the folded magazine together to form the clown's body.

2. FEET: Cut the feet from cardboard and glue to the bottom of the body.

3. Spray the body, feet and 1½'' white foam ball balloon with blue paint. Allow the paint to dry.

4. NECK RUFFLE: Cut two 5'' wide, yard long pieces of net and gather them into a bundle in your hand. Place a doubled rubber band around the center of it to hold it together. Fold the ruffle down around the band.

PATTERN FOR FEET
To make a full size pattern for the feet, trace around one foot onto a sheet of folded paper. Cut and then open into a full size pattern.

5. HEAD: Stuff scrap white foam into the head. Place the rubber band of the ruffle around the neck and arrange the ruffle. Attach the head to the body with a doubled piece of chenille stem.

7. SLEEVE: Cut a 2'' wide, two yard long strip of net. Accordion pleat the strip into 2'' folds. Round off the corners as shown. Make a hole thru the center of each circle. Push the chenille arm thru the hole and pull the sleeve into place as shown. Repeat for the other sleeve.

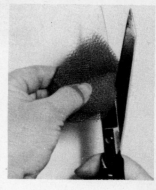

6. HAND: Cut an 8'' length of blue chenille stem for an arm. Form one end into an 1'' loop and then shape into a hand as shown below. Repeat for the other arm and hand.

8. See photo on opposite page and add round circle trims of white chenille and red bumps down the front of the clown and on each toe.

9. BALLOON: Push an 8'' piece of 18 ga. wire into foam ball and fasten into clown's hand.

PATTERNS FOR FLO, THE FISH --- See page 88

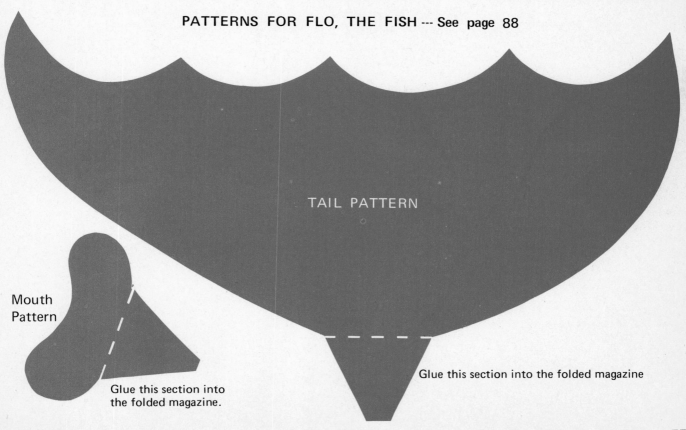

TAIL PATTERN

Mouth Pattern

Glue this section into the folded magazine.

Glue this section into the folded magazine

Noel Trimmin' Fun

Gay gold bells and a red caroler speak of traditional Christmas trim. You will want to make these by turning to page 86 for the caroler and pages 82 and 83 for the bells.

Three Kings

These stunning Three Kings were made by folding 11" magazines. By combining two different folds, you can get an unique effect of a gown in front with a cape around the shoulders and across the back. See the next page for easy-to-follow instructions. Gold paper lace trim gives an elegant old-world touch while gay jewels and sequins are added for sparkle.

Three Kings

Proudly they followed the guiding star, bearing their noble gifts and good tidings --- composing a beautiful holiday tableau for your home. All three wisemen are luxuriously clad in jewel-bedecked "robes" and carry sparkling treasures for gifting. Special attention has been devoted to the many fancy trims like gold paper lace details, which are so easy to apply and add superb, rich accent to the colorful robes. Each of the three kings stands 15" tall. You'll cherish this trio year after year.

MATERIALS NEEDED FOR EACH

8¼" x 11" magazine , 226 pages of more
2 Craft spray paints, colors of your choice
2 Jumbo white chenille stems, arms
4" dia. Foam ball, head   White craft glue
4" dia. Gold paper lace medallion, collar
2 pcs. 6" x 6" White paper, sleeves
Chenille bumps, white and black; whiskers
1" x 1" black paper, eyes
Gold paper lace trim   Jewels and sequins
Scrap material for headdress

DIRECTIONS

1. Remove the front and back covers from the magazine. The first fold for all the pages is the same as the first fold for the Graduate on page 71.
2. See the diagram at the top of the next page and make a cardboard template. Fold the first eighty pages around it, one at a time.
3. The second and third fold for the next eighty pages is the same as the second and third fold for the Graduate as shown on page 71
4. Fold the remaining pages as in step 2 above.
5. Glue the front and back pages together and stand upright to form a body shape as shown in Fig. A. The effect simulates a gown with a cape. If a fuller body is desired, fold a second magazine and glue it to the first magazine.
6. See Fig. B and spray paint the front of the gown one color and the cape another color.
7. For each sleeve roll and glue a 6" x 6" piece of paper into a cone shape sleeve and spray paint them.
8. Glue a 10" length of jumbo stem chenille into the sleeve for an arm. Roll the other end into a hand. Repeat for the other arm. Glue into place.

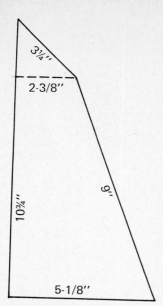

Diagram for cardboard
template to fold around.

Fig. A   Folded magazine body
simulates a gown and cape.

Fig. B   Protect balance of pages with a
piece of cardboard while spraying.

9. Place a 4'' diameter gold paper lace medallion collar over the top of the body.

10. Double a 9'' length of 18 ga. wire. Insert one end into the 4'' foam ball head. Glue the other end into the hole at the top of the body, going thru the collar.

11. Glue gold paper lace trim, jewels and sequins to the gown as desired.

12. Add a cloth headdress, crown, face features and chenille bump whiskers of your choice --- see photo.

EYE PATTERN
Cut of black paper

Ralph...
the Moose

Here's an idea for a fun animal; Ralph, the Moose.

A Reader's Digest is folded the same as for the Graduate on page 72 A 4'' foam ball forms the head. The nose is carved from a piece of foam and attached to the head with 2'' lengths of chenille.

The 5''x 3¾'' antlers are cut of cardboard and pushed into slits cut in the head. The moose is then sprayed with antique gold paint.

Comical 18 mm moving eyes and red yarn hair add the finishing touch to Ralph.

Have fun making him!

Contemporary
Bell Hanging

Although these bells are whisper quiet, they pay a brilliant tribute to the Christmas Season. The rich gold-colored bells feature three interesting shapes --- created with three different sized magazines. Dainty gold paper lace trims, bows and a yarn tassel add a "note" of charm. The 6 to 7-ft. long hanging is a perfect holiday dress-up for living rooms or hallways. Or, choose electrifying colors for the bells for a ring-a-ding-ding luau decoration!

MATERIALS

Pocket-size paperback book
8¼"x 11" magazine
2 Telephone directories,
    approx. 400 pages each
12"x 28" heavy cardboard
White tacky craft glue

Assorted gold paper lace trim
16 ga. wire, covered
6 Lg. ribbon bows, green
6 Clusters of ball ornaments,
7" yarn tassel, red
Gold spray paint

FOLDING DIRECTIONS; first remove the covers.

FIRST BELL, at the top, is a pocket-sized paperback book. Fold each page as shown in Fig. A, step 1 and step 2 below.

SECOND BELL, from the top, is an 8¼"x 11" magazine. Fold each page as shown in Fig. A, step 1 and step 2 below.

THIRD BELL, from the top, is a phone directory. There are two different folds used. Each fold is made alternately on every other page. The first page is folded like Fig. A, step 1 and step 2. The alternate page is folded lengthwise two times as shown in Fig. B, step 1 and step 2 below.

FOURTH BELL, from the top, is a phone directory. Fold as in Fig. A, step 1.

FIFTH BELL, from the top --- or bottom bell, is a Reader's Digest. Fold each page as in Fig. A, step 1.

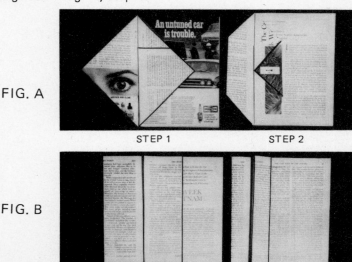

FIG. A   STEP 1   STEP 2

FIG. B   STEP 1   STEP 2

ASSEMBLY DIRECTIONS

1. Glue a wire to each bell so they can be connected together. To do this cut a piece of 16 ga. wire 4'' longer than the binding of each magazine. Bend a hook in each end of the wire. Glue the wire to the book binding. Locate the bell in the center of the wire and apply tacky glue.

2. Cut a cardboard backing for each bell and glue to the backside so the bell is in a half round shape.

3. Completely cover the bells and cardboard backing with gold spray paint. Allow the paint to dry.

4. Connect the bells together by securing a piece of 16 ga. wire to the hooks. Allow approximately 4'' between each bell and add a ribbon bow and ball ornaments in this space.

5. Glue assorted gold paper lace trim to the bells as desired.

6. Add a yarn tassel to the bottom hook and your hanging is ready to add an exotic accent to your wall, hallway, along the mantle or wherever it is hung. □

Bell Plaque

MATERIALS

2 Reader's Digest
2 pcs. 7½''x 7½'' cardboard
Gold spray paint
12''x 36'' White foam plaque
Gold paper lace trim
5 yds. Ribbon, green
Stemmed flocked balls, red
White tacky craft glue

DIRECTIONS

1. Remove the front cover from a Reader's Digest. Cut down thru the binding leaving ninety pages fastened together. You will use these pages.
2. The first fold for each page is like the first fold for the Graduate on page 71
3. See the diagram at the right and make a cardboard template. Fold each page around it.
4. Cut a cardboard backing and glue the half round bell to it. Repeat for the other bell.
5. Spray paint the bells gold. Allow the paint to dry and glue stemmed ball in place for a clapper.
6. Pin bells in position on foam plaque.
7. Pin ribbon bows and flocked balls to plaque. Pin gold paper lace trim around edge of plaque.

2½''
1¾''
5''
3-5/8''
1-3/8''
3-1/8''

Diagram for cardboard template to fold around.

Bells were made for hanging --- on a special plaque like this! And, these lovely bells are still another version of the galaxy of shapes you can make with the Reader's Digest magazine. A swirling green satin ribbon links the charming chimers.

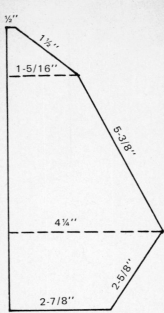

DIAGRAM TEMPLATE
Make a cardboard template to fold around.

Template measurements: ½", 1½", 1-5/16", 5-3/8", 4¼", 2-5/8", 2-7/8"

EYE PATTERN
Cut 2 of felt

Gordon the Gobbler

Let's talk turkey about Gordon the Gobbler --- the center of attention at your Thanksgiving feast. You can make Gordon in a jiffy and keep him for your table, mantle, or console year after year. Gordon's body is a bright brown color and his foam head is accented with wiggle eyes. To make this gobbler a very proud bird, display Gordon with a floral arrangement or gay wax candle.

MATERIALS
Reader's Digest magazine
1-7/8" x 2¾" White foam egg
2 Chenille bumps, red
4 Chenille stems, lt. brown
1 Jumbo chenille stem, dk. brown
Pr. of moving eyes, 8 mm
Madras tissue paper, orange/brown
Spray paint, brown
1" x 1" Felt, lt. green
2 Pcs. wire, 18 ga.
Green floral tape
8" x 12" White foam block, green

DIRECTIONS
1. BODY: Remove the front and back covers from a Reader's Digest. See Fig. A for the first fold.
2. Make a template, as shown above, and fold each page two more times. See Fig. B on page 85. Glue each page at X.
3. Spray paint the folded magazine body and foam egg head with brown paint. Allow the paint to dry.
4. NECK: Connect the head to the body by folding the brown jumbo stem double. Insert and glue one end into the hole in the body. Shape the neck and glue the other end into the bottom of the egg head. The neck is about 4" long.
5. EYES: See pattern and cut eyes of light green felt. Glue onto the head. Then glue the moving eyes into place.

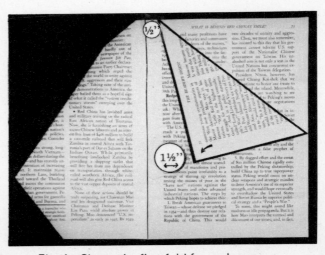

Fig. A Shows the first fold for each page.

84

6. BEAK: Fold a 1½" length of light brown chenille stem in half to form a beak. Insert it into the pointed end of the egg head.

7. Glue the red bump "gobbles" in place.

8. LEGS: Wrap two 4" lengths of 18 ga. wire together with floral tape. Wind a light brown chenille stem around the wire to cover it. Leave 1" bare at the bottom; this will be stuck into the foam base later. Repeat for the other leg. Glue end of each chenille-wrapped wire into the bottom of the turkey's body. See Fig. C and notice the position for the legs.

9. FOOT: Bend a light brown chenille stem into a foot as shown in Fig. D. Repeat for the other foot. Slip a foot onto the bottom of each leg and then push the legs into the foam base so the turkey stands upright.

10. TAIL: Cut a 20"x 30" sheet of Madras tissue into three pieces each 20"x 10". Glue them together to form a 60"x 10" strip. This is done so the color stripes run in the right direction for the tail. To make the fan-shaped tail, accordion pleat this strip into 1" pleats. See Fig. E and cut one end off at a 90 degree angle.

11. Pin the colorful tail to the back of the turkey so it fans out like a real turkey tail. □

Fig. B Shows folding sequence. Add glue at spot marked X.

Fig. C Glue legs into the bottom of the turkey.

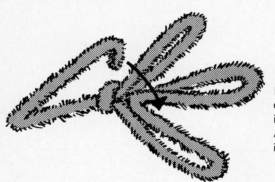

Fig. D Fold a chenille stem into a foot. Actual size is shown.

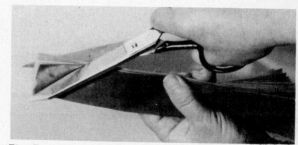

Fig. E Cut off one end of the tail on a slant.

Sue Belle, the Caroler

Sounds of ringing bells, pungent scents of pines and jolly smiles of the department store Santas tell you happy holidays time is here! Now, add to your traditional annual decoration cheer with Sue Belle the Caroler. She's an easy-to-make miss --- dressed in scrumptious red with paper doilies adding a white lacy frosting to her gown. There's extra eye-catching artistry in the lovely tiny paper roses that trim the hem of her skirt.

MATERIALS
Reader's Digest magazine
Doll head, 4'' dia., stemmed
Diamond dust
Paper doily
White craft glue
2¾'' x 4¾'' gold deco-foil
2 Jumbo chenille stems, 9 mm, white
4 sht. 20'' x 30'' tissue paper, any color
10½'' dia. lightweight cardboard disc
Spray paint, color of your choice

Sue Belle (who would guess her gown was once a Reader's Digest) is shown in color on page 67. She stands approximately 12'' high and her skirt is 11'' across.

Fig. A Staple each fold close to the center of the magazine.

DIRECTIONS

1. Remove the front cover from a Reader's Digest magazine. Cut down through the binding, leaving eighty pages fastened together. These pages will be folded into a pleated skirt.

2. You will fold two pages at a time. DO NOT crease the folds. Fold one set of double pages so they face another set of double pages. Each fold will be stapled as close to the center of the book as possible. See Fig. A. This insures a rolled pleat effect.

3. Continue until all doubled pages are folded and stapled. To keep the pages from creasing as you work, allow the folded pages to hang over the edge of a table so there is no pressure on

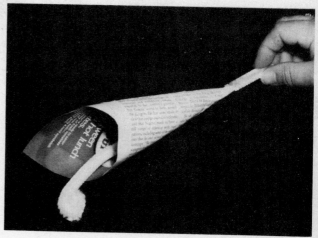

Fig. C Shows sleeve with chenille arm and hand.

them. To form the skirt, stand the book up and it will open into a circle as shown in Fig. B below.

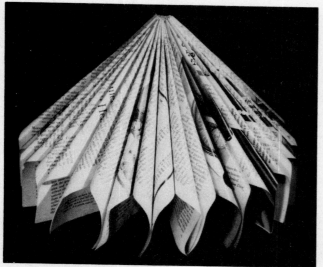

Fig. B Stand the book up so it opens into a round, pleated skirt.

4. Glue the circular cardboard disc to the bottom of the doll's skirt.

5. SLEEVE AND ARM: Cut 2" off the top of a page. Roll and glue the remaining page into a cone shape sleeve. Insert a jumbo chenille stem for an arm. Allow the stem to extend 1" at the top. Roll the other end of the stem into a hand. See Fig. C.

6. Secure sleeve by gluing arm stem, at top of sleeve, into top of skirt. Repeat for other sleeve.

7. ROSES: (You will need 23.) Cut a 3"x 30" strip of tissue paper for each rose. Fold the strip lengthwise and glue the edges together.

8. Run a bead of glue along the bottom edge of the glued strip and begin rolling to form a rose. Gather the bottom as you roll so the rose is open and full at the top. See Fig. D. Pinch the glued bottom together.

Fig. D Roll tissue strip into a rose.

9. Glue a rose into each of the twenty openings around the bottom of the skirt. Glue three small roses down the front of the skirt as shown in the photo on page 86.

10. Spray paint the skirt, roses, sleeves and chenille arms. Sprinkle diamond dust on the wet paint for sparkle.

11. Insert the stem of the head into the hole at the top of the skirt and glue to hold. Glue a paper doily trim around the neck and inside each sleeve.

12. Fold the 2¾"x 4¾" gold deco-foil in half for a song book and glue into the caroler's hands. ☐

Flo, the Fish

If you've been fishing for a novel decoration for your patio or need an interesting variation for a luau decor --- don't let this one get away! Flo the Fish will cause ripples of excitement with her exotic colored body and friendly moving eyes. Just follow the easy instructions and create this prize catch. But don't stop with Flo --- why not make a whole school of vibrantly colored fish? Use a fisherman's net as a backdrop, add some of your favorite shells and your tropical fish collection, and you'll have an "island paradise" in your own home.

FIN

MATERIALS
Reader's Digest magazine
6"x 7½" lightweight cardboard
5"x 5" lightweight cardboard
1 Moving eye, 18 mm
Spray paint, pink or color of
      your choice
Poster paints, black and yellow
White craft glue

DIRECTIONS
1. Remove the front and back covers from a Reader's Digest and fold each page into a two-fold as shown in Fig. A below.
2. Since the backside of the folded magazine body is flat, glue a 5"x 5" piece of cardboard to it.
3. Cut mouth, fin and tail of cardboard and glue them into place. See page 77 for tail and mouth patterns.
4. Spray paint the fish and allow the paint to dry.
5. Glue eye into place.
6. Hand paint mouth black. Paint trim on body, tail and fin as shown in the photo above.

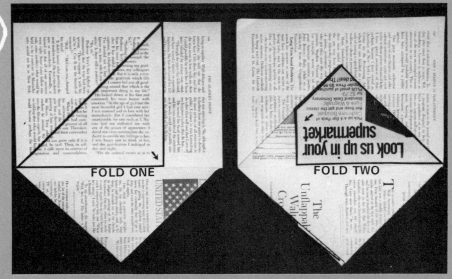

FOLD ONE   **FOLD TWO**

Fig. A   Fold each page into this two-fold sequence for the body of the fish.

PLASTIC MASQUERADE
USING THROW-AWAY BOTTLES AND LIDS

A CRAFT COURSE BOOK

COME TO THE MASQUERADE

Come to the Masquerade! Meet Miss Darling, the Easter Duckie who never grows old. Dance a hoe-down with Granny Annalee, a former liquid soap bottle turned sewing caddy. Revel in the rainbow colors of plastic milk bottle patio lights shining with thankfulness for their salvation from the trash barrel. Then surrender to the charms of an exotic beaded curtain of Shrinky Lids. (Their secret rests in the inner sanctum of this book.) Do it all at the Plastic Masquerade Ball, as you waltz through these pages of exciting plastic projects designed to disguise the hum-drum in a cloak of, well — fun!

But come prepared, lest you turn into a pumpkin at the stroke of midnight! Follow these general instructions when dealing with plastic containers. They'll help you change that pedestrian plastic into a magical masquerade of plastic personalities!

1. Work with clean containers. Rinse the inside thoroughly. Soak the outside to remove labels. Clean off glue with acetone (nail polish remover) or lacquer thinner.

2. Ease your cutting chores by first warming the container. Placing it in sunlight or filling it with hot water will soften the plastic. Use a sharp craft knife, single-edge razor blade, or scissors to cut the plastic.

3. Take advantage of the brightly colored bottles on the market. Or use permanent, felt-tipped marking pens to give lasting color to plastic containers. A quick spray of clear acrylic followed by acrylic spray paint will also do as a cover-up.

Then let your imagination coax those coy containers into taking off their commonplace plastic masks — and appearing as their celebrated alter-egos at the Plastic Masquerade!

SAVE THE FOLLOWING PLASTIC THROW-AWAYS FOR MAKING THE PROJECTS IN THIS BOOK.

* Powdered cleanser containers
* Liquid detergent bottles
* Fabric softener bottles
* Milk and juice bottles
* Liquid bleach bottles
* Lids and meat trays
* Whip cream tubs
* Legg's eggs

Jardiniere

An empty plastic container need not become a throw-away nuisance. Instead, it can be cut and woven with colorful yarns into an attractive jardiniere for a favorite plant. All you'll need is a bright-colored 17 oz. powdered cleanser container, yarn in assorted colors, a 1/8" paper punch, and a tapestry needle.

After making this jardiniere, you'll be eager to experiment with other woven projects such as place mats, tumbler jackets, and doilies. The secret is to cut an UNEVEN number of strips!

DIRECTIONS

1. Cut off the top of the container 4½" above the base.

2. Cut the container into an UNEVEN number of ¼" strips. Leave ¾" of plastic above the base uncut. Punch a hole ¼" down from the end of each strip.

3. To weave, tape one end of yarn to the inside bottom of the container. Starting at the bottom, weave the yarn over every other strip. Knot on the inside each time you change colors. Weave up to the holes.

4. To finish the edge thread a needle with yarn; blanket stitch around the top making two stitches in each hole. □

ANNALEE

MATERIALS

*Plastic detergent bottle
Plastic doll head, 1½" dia.
Curly angel hair
2 pcs. art foam, 3¼"x ¾" ea.
White craft glue
Sand or casting plaster
Plastic hands

Six 12" pcs. wire, 22 ga.
Two 12" pcs. wire, 28 ga.
Acrylic spray paint, gold
¾ yds. red gingham
12"x 6½" red percale
2 yds. black lace, 5/8" width
3"x 7¼" interfacing

*22 fl. oz. size — Bottles with an hourglass shape, similar to those shown above, are recommended.

DIRECTIONS

1. Spray paint the bottom 2" of the plastic bottle. See Fig. A.

ARMATURE

2. ARMS: Heat an ice pick or an awl and pierce an armhole thru each side of the plastic bottle about 1¼" from the top. Twist six pcs. of 22 ga. wire together for the arms. Run the wire thru the armholes. Sew a piece of art foam padding around each arm. Bend the bottom ½" of each arm at a right angle; glue a hand on the bent part. See Fig. B. Bend the arms into shape.

3. HEAD: Shape eyeglasses of fine 28 ga. wire and attach them to the plastic head. Glue curly angel hair around the front top portion of the face. See Fig. C.

4. Fill the bottom 2" of the plastic bottle with sand or casting plaster.

5. Glue the head to the bottle cap and screw the cap into place. See Fig. D, on page 92 for the complete armature.

CLOTHES (See pattern on page 21.)

6. BLOUSE: Hem each sleeve ¼". Topstitch lace onto the sleeves. With right sides together, stitch up the sides and sleeves of the blouse. Turn right sides out.

7. BIB: Hem the outside curved edge of the bib ¼". Topstitch an 18" length of gathered lace over the hem. Fold over ¼" of the open left edge and press to stay.

Granny Annalee looks every inch the seamstress in her homemade, red gingham frock with matching bonnet. Enlist her as your helping hand on sewing days. You'll agree she's as functional as she is cute when you find you can store needle, thread, measuring tape, and thimble in her perky, red percale pockets. Your stitches will always be in time, when Granny keeps your notions in line!

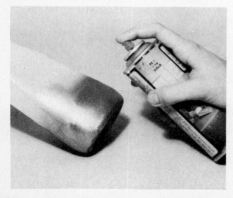

Fig. A Spray paint the bottom 2" of the bottle.

Fig. B Glue a hand on the bent part of the wire arm.

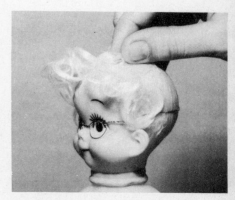

Fig. C Glue hair around the front top portion of the face.

91

Fig. D   The complete armature.

8. SKIRT: Make a ¼" hem along one 22" edge of gingham. Gather the other 22" edge to 7". Cut a 9" x 1" strip of gingham for a waistband. Place the strip over the gathering with right sides together, leaving 1" extended on each side of strip. Keep the top edges flush. Stitch. Fold the strip over the gathering and topstitch thru all thicknesses. Stitch the skirt together in back, leaving 2" open at the top.

9. APRON: Cut a 12½" x 7" pc. of percale. Gather one of the 12½" ends to 3½". Cut a 12½" x 1" pc. of gingham and stitch this to the bottom edge of the apron just as you did the waistband for the skirt, this time placing the right side of the gingham on the wrong side of the apron. 2½" below the waist, clip each side edge of the apron. Fold over ¼" of edge between the cut and the waistband with the wrong sides together. Hem. Fold over ¼" of apron edge below the cut, with right sides together. Hem. Fold up 2¼" of the apron bottom, right sides together, and topstitch the sides and along the hemline. With topstitching, divide the folded portion into seven pockets. Center and topstitch a 21½" pc. of lace over the waist so the ornamental edge is turned up.

10. CHOKER: Fold a 6½" x 1½" pc. of gingham into thirds. Press to stay.

11. BONNET: Cut a 7½" circle of gingham. Stitch around edge and gather for a cap. The cap should fit around the back half of the head behind the ears. Ease gathering if necessary. Cut two pcs. of gingham and a pc. of interfacing for the bonnet rim. Place the two pieces of gingham on top of each other, right sides together, and place the interfacing on top of this. Stitch thru all thicknesses around curved part of rim. Turn right side out. Topstitch lace on curved part of rim. Topstitch a quilting pattern on rim if desired. Place unstitched end of rim on gathered edge of circle, right sides together. Stitch thru all thicknesses. Cut two 6½" pcs. of lace. Hand-stitch each to the rim of the bonnet for bonnet strings.

12. Dress the doll. Secure open ends of clothing with pins.   □

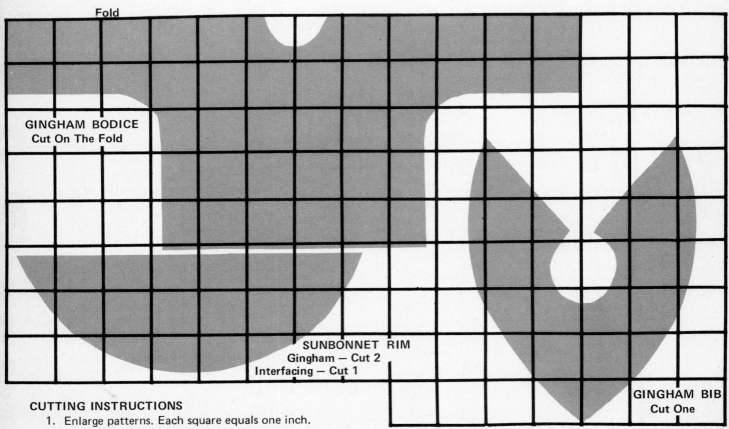

Fold

GINGHAM BODICE
Cut On The Fold

SUNBONNET RIM
Gingham — Cut 2
Interfacing — Cut 1

GINGHAM BIB
Cut One

CUTTING INSTRUCTIONS
1. Enlarge patterns. Each square equals one inch.
2. Cut out material as instructed on pattern.
3. Cut additional pieces listed below:
GINGHAM APRON TRIM: Cut a 12½" x 1" piece.   GINGHAM WAISTBAND: Cut a 9" x 1" strip.
GINGHAM CHOKER: Cut a 6½" x 1½" piece.   GINGHAM SUNBONNET CAP: Cut a 7½" circle.
GINGHAM SKIRT: Cut a 22" x 6¼" piece.   PERCALE APRON: Cut a 12½" x 7" piece.

MISS DARLING

"Tisket-a-Tasket, I'm an Easter Basket"
Miss Darling is shown in color on page 102

MATERIALS

1 gal. bleach bottle, white
3" dia. white foam ball
1 pc. ea. 9"x 12" felt;
  yellow and orange
¾"x ¾" felt, black
8 chenille stems, white

3 chenille bumps, yellow
½ yd. lace, 5/8" width
½ yd. rick rack, white
9'x 12' thin, plastic drop cloth
White, thick craft glue
Optional: small flowers and
  feathers for trim

DIRECTIONS

1. Double thick felt is needed for stiffness. Cut the 9"x 12" yellow felt in half crosswise. Glue the two 6"x 9" pieces together by first coating one piece completely with white glue. Cut the orange felt in half lengthwise and glue the two 4½"x 12" pieces together.

2. Make tracings of the overlay patterns. Cut one yellow beak and two yellow feet. Cut one orange bonnet. Cut the ¾"x ¾" black felt on the diagonal for eyes.

3. Gather and glue the lace around the curved rim of the bonnet. Center and glue the rick rack along the straight edge; the ends will become bonnet ties.

4. Fold the beak in half and pin to the foam head. Glue the eyes in place. Pin the bonnet in position. Cut the chenille bumps apart and twist each into a curl; glue these to the front top of the head. Set the head aside until step 10.

5. With a sharp craft knife or scissors cut an opening in the front of the bottle. Start 1½" up from the bottom and make the opening 4¾" high and 4½" wide.

6. With a pointed tool punch sixteen holes around the opening. These should be ¼" in from the edge and about 1¼" apart. See Fig. A.

7. Refold the drop cloth over and over so it's 4" wide and 12' long. This will be several layers thick. To make a tuft cut off a 4" width and open it into a 4" wide strip, 12' long. Loosely wind this strip around a piece of 5"x 2" cardboard; wrap around the 2" width. Slip a 6" length of chenille stem under the strip and twist the stem ends together. Cut thru the strip at the opposite fold. See Fig. B. Fluff the tuft and trim the edges even. Make sixteen tufts.

8. Staple the feet to the bottom edge of the opening. See Fig. C.

9. To attach tufts insert the stem of a tuft thru each hole around the opening. Spread the stems open on the inside so they won't pull out.

10. Cut a hole in the bottom of the foam head and glue the head onto the top of the bottle.

11. Cross and pin the bonnet ties under the chin. Add any decorative trim desired, such as flowers and feathers. □

Look what the Easter Bunny left for the wee egg hunters in your household — an Easter Basket duckie that's more fun than a basket full of Easter ducklings! Imagine the excitement of a toddler who reaches through the soft, plastic-tuft-lined rim of this basket to find eggs on a bed of easter grass, and maybe a chocolate bunny or two. Miss Darling Duck guards them from April showers and inquisitive garden bugs in her plastic bleach bottle body.

Fig. A Punch holes around opening.

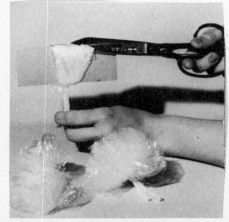

Fig. B Cut thru strip at fold.

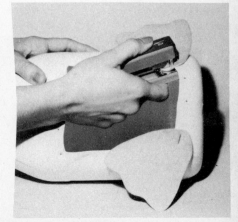

Fig. C Staple feet to bottle.

The colorful "pinkie" butterfly ring and apple blossom charm bracelet are fashioned of shrinky cut-outs.

SHRINKY LIDS

QUESTION: *"What's made of polystyrene plastic, shrinks and thickens when heated in a low oven, and becomes a 2" oval medallion?"*

ANSWER: *"Shrinky Lids"!*

They're the shrinkable, versatile, and totally likable lids of the lowly liver containers* found at your favorite meat counter or hobby shop. Color them with permanent marking pens, shrink them into medallions or cut-outs, then fashion them into:

charm bracelets — "pinkie" rings — earrings — key fobs — buttons — lavalieres — beaded curtains — and wind chimes.

*Some containers may hold oysters, or what-have-you, instead of liver!

MATERIALS
Plastic liver lids
Permanent black felt-tip marking pen
with a fine point
Permanent felt-tip marking pens in
assorted colors

Please read the TIMELY TIPS, on the opposite page, before beginning a "shrinky" project. See pages 98 and 99 for additional designs and projects. Then turn to page 96 for projects in "living color".

DIRECTIONS
1. Place a plastic lid over a pattern and trace the design onto the lid with a fine-point, black felt-tip pen. Color in the design with permanent felt-tip pens.
2. If a cut-out shape is desired, rather than an oval, cut around the design with scissors. Make a hole in the top with a paper punch.
3. Pre-heat the oven to 325 degrees. DO NOT exceed this temperature. Place the decorated lid on a foil-covered cookie sheet and put in the oven for one to three minutes. The plastic curls and goes thru all kinds of contortions as it shrinks but will finally straighten out.
4. Remove from the oven when the medallion is flat and approximately 2" in diameter.
5. Optional: A cut-out can be bent into shape before the plastic cools. Re-heat for additional shaping, as needed. This method is suggested for shaping the butterfly wings.
6. If a hole is desired in the medallion, drill one with a craft motor tool or pierce one with a heated awl or ice-pick.
7. A jump ring can be fastened thru the hole. See Fig. A on page 95 The jump ring is the connecting link between the medallion and any jewelry finding used. □

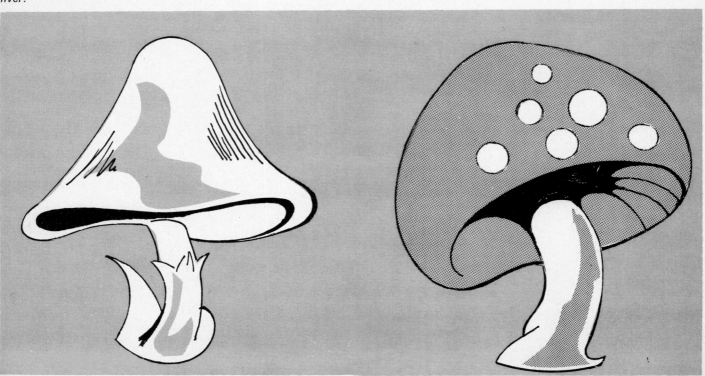

Fig. A Secure a jump ring to the medallion.

When only half of a symmetrical pattern is shown, trace a full-size pattern by placing the straight edge on the fold.

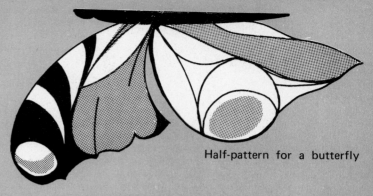

Half-pattern for a butterfly

TIMELY TIPS

1. COLOR: Only PERMANENT felt-tip pens are recommended; other coloring media seem incompatible with this method. The design shrinks in proportion with the lid causing the colors to intensify, so experiment with color density. Apply colors with long, even strokes. One color can be shaded over another dried color.

2. CUT-OUTS: For interesting shapes cut out a design with scissors prior to shrinking it. Cut leaves with pinking shears for a serrated edge.

Cut the rim off of a lid, color and shrink it for an unusual twisted bauble. Some are included on the wind chimes shown on page 99.

3. HOLES: These are needed for jump rings and can be made with a paper punch before heating a cut-out design. Otherwise drill holes in a finished medallion with a motor tool or pierce with a hot awl. A round lid will shrink into a slight oval shape, unpredictable in direction. This may determine when and how you make the holes.

4. SHAPING: Cut-outs, like a butterfly, can be given dimension by shaping the wings while they are warm and pliable. If the plastic cools and hardens, re-heat for additional shaping. A medallion can be flattened while still warm.

5. OTHER SHRINKABLE MATERIALS: The corrugated and puckered surfaces of most clear plastic meat and cookie trays have repetitive designs which can be colored before shrinking. Assorted meat tray medallions are featured on the wind chimes on page 99. Clear plastic cups shrink into interesting shapes which are fun to use as reflectors for Christmas tree lights or package tie-ons. ▫▫

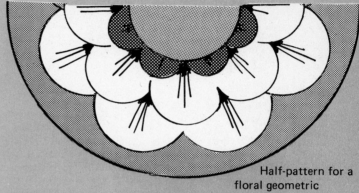

Half-pattern for a floral geometric

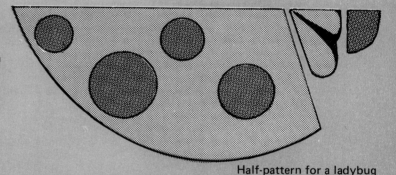

Half-pattern for a ladybug

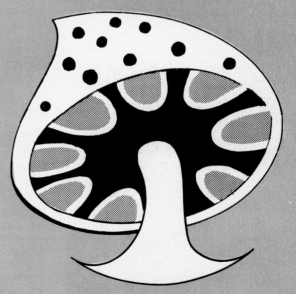

BEADED CURTAIN

Divide and conquer a room with a lovely beaded curtain accented with 2" hand-decorated medallions.

This decorative drape was designed to inspire creativity with "shrinkables" - - - or turn to page 8 for a collection of charming medallion designs to use on liver lids. The colors can be of your own choosing.

Create a dramatic hand-fashioned bead curtain to add a "new mood" to your favorite room today.

COLORFUL "SHRINKABLES"

TRIFLES AND TREASURES

Make a special gift for a special person!

An apple blossom charm bracelet to delight a teen-ager - - - a mushroom medallion key fob to keep keys within reach - - - butterfly earrings to enhance milady - - - a snail magnetic to hold notes on metal - - - and cut-out mushroom pins - - - All are fashioned of shrinkable lids.

FOR THE DESIGNING LADY ---

With a fascinating selection of doll heads and face masks available at your favorite hobby shop, you can have the fun of originating your own dolls. Simply make an armature similar to Annalee, on page 91, and dress it as desired. The Jamaican and Granny dolls were made on 32 fluid ounce plastic detergent bottles; their hands are of felt. These two dolls are shown as ideas to stimulate your designing instincts.

The colorful patio lights are easy to make by following the instructions on page 101. They promise to add a gay, festive touch to your garden night-life.

BEADED CURTAIN

The six interesting designs on this page were used to decorate the liver-lid medallions featured in the stunning curtain shown on page 96.

To assemble the curtain, jump rings were placed in holes drilled thru the top and bottom of the shrunken medallions. Then the medallions were strung with short lengths of colorful plastic beads between them. A 1" brass ring was secured to the top of each string. See Fig. A. To hang the curtain, simply run a rod thru the top rings. A plain string of fused plastic beads was hung between each decorative string. The medallions and beads should be in colors to match your decor.

Approximately nine medallions are needed on each 32" length string. □

Fig. A Illustration of decorative string assembly.

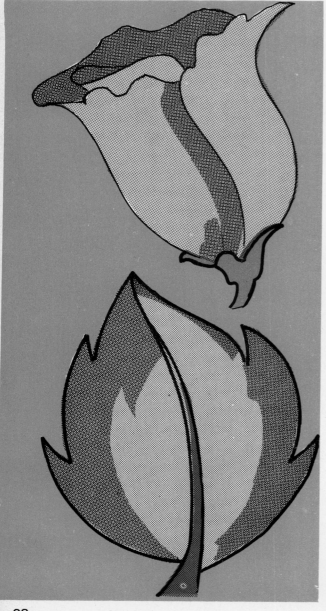

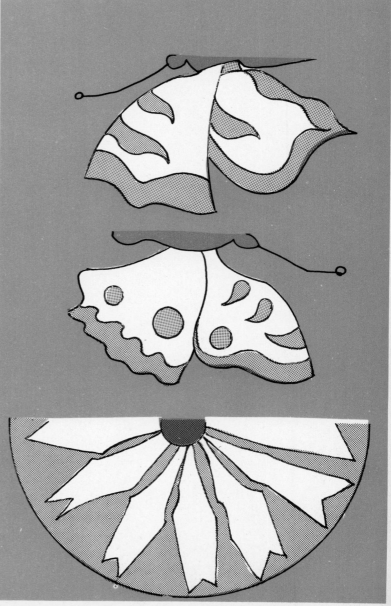

Cut 3 leaves from one lid

WIND CHIMES

The merry tinkle of plastic being caressed by a gentle breeze, creates a delightful symphony all its own.

These charming chimes were fashioned of shrinkable plastic. You will notice that the raised corrugated patterns on meat trays, highlighted with color, resulted in circular designs on rectangular medallions. Coloring the puckered surface of a cookie tray, produced colored squares. The owl and mushroom designs drawn on the round liver lids, were traced from patterns on pages 95 and 99. The twisted baubles were formed by cutting off and coloring the rims of liver lids.

This wind chime was easily made by assembling the medallions with jump rings. Strings of chimes were then tied to a weathered twig with colored cord.

Have the fun of making one and then enjoy the lyrics of the wind. □

WIND FLOWER

MATERIALS

8 plastic panels from fabric
    softener bottles,
    4"x 8" ea.; white or
    yellow
Heavy plastic drinking straw
Acrylic spray paint;
    orange and green

3" dia. foam ball
3/8" dia. dowel stick
Wire coat hanger
14 mm. wooden bead
Masking tape
Tacky white glue

INSTRUCTIONS

1. Trace and cut a petal pattern of lightweight cardboard. Cut 8 plastic petals.

2. Make a hole thru the center of the foam ball with a pencil. Cut a 3¾" length of plastic straw. Coat this straw with glue and push it thru the hole in the ball until one end of the straw is flush with the ball at the front and the other end extends ¾" at the back. Allow the glue to dry.

3. Wrap a piece of masking tape around the circumference of the ball as a guide for the placement of the petals. See Fig. A.

4. Pull back about 1" of tape. Bend a petal lengthwise and force the tapered end into the ball where the tape was. Position it at a slight angle. See Fig. B. Continue adding the other petals, placing each one so the center fold is about 1¼" from the center of the petal next to it.

5. Remove the petals and spray the ball and straw orange.

6. Drill a small hole thru the dowel stick ½" from the top. Drill another hole ½" below it. Spray the stick green.

7. Cut a 6" piece of wire from the hanger. Bend one end into a 1" hook. Fit the hooked wire into the holes of the dowel stick, placing the 5" length thru the top hole.

8. With the protruding section of straw facing the stick, slide the straw and ball over the 5" length of wire. Leave 1/8" between the stick and the straw. Glue a bead over the wire extending out of the front of the flower. This will keep the flower from sliding off the wire. Allow the glue to dry. Clip the wire flush with the bead.

9. Replace the petals by gluing the tapered ends of the petals into the slots around the ball. □

A garden is a bright and happy place --- and this gay wind flower will be very much at home tucked among the flowers and greenery. She will add her own colorful charm as she whirls to any tune a breeze will blow for her.

Your wind flower could be a cheerful yellow, snow white, or even blue depending on the color of plastic bottles you've been saving. Turn to page 102 to see a bright yellow one.

This showy whirligig measures 16" from petal tip to petal tip --- and has real eye-catching appeal!

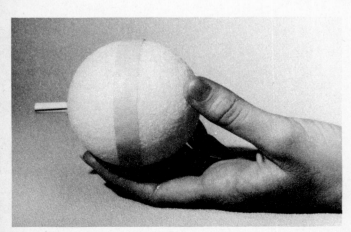

Fig. A   Guide tape in place.

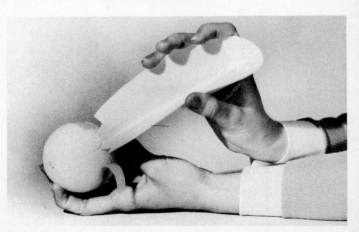

Fig. B   Positioning a petal.

PATIO LIGHTS

Bright nights begin with gay lights so set the mood by shining a little color around your garden or patio.

A set of ordinary outdoor Christmas lights will suddenly take on a new glow and charm when you add "frosted" shades cut from plastic bottles.

The lovely muted colors, so simply produced, are sure to add an air of excitement to your outdoor wonderland — be it a summer evening or the night before Christmas!

MATERIALS
Set of outdoors Christmas lights
Quart-size plastic juice or milk bottle; one for each light.
These should be translucent and without printing.

DIRECTIONS
1.
a full-size paper pattern; make any necessary adjustments so the pattern fits the bottle being used. A point should come at each curved corner.

2. Tape the pattern around the bottle as shown in Fig. A. Trace around the pattern with a felt pen.

3. Cut off and discard the neck of the bottle. Cut along the pattern lines with a pair of scissors. See Fig. B.

4. Use a craft knife and cut a ¾" diameter hole in the center bottom of the bottle; the bottom of the bottle will become the top of the shade.

5. To connect the shade, hold a light socket against the hole on the outside of the shade; place a bulb inside the shade and screw it into the socket. See Fig. C.

6. Make a shade for each light bulb on the string. You can use various colored lights or all one color, depending on the effect you want to achieve. □

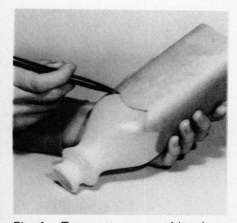

Fig. A   Tape pattern around bottle.

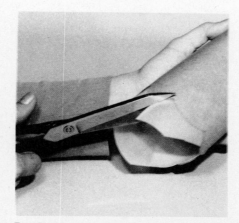

Fig. B   Use scissors to cut out shade.

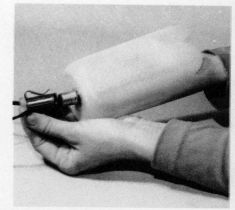

Fig. C   Connect shade to string of lights.

WIND-BLOWN DAISY

For a flair-of-fun, make this gay wind flower to add a little color to your garden. It promises to bring back nostalgic memories of the days when you were fascinated with the whirling of a pinwheel held up to a breeze.

The young-in-heart - - - and who isn't! - - - will continue to be delighted whenever a wind turns this daisy's head.

Instructions can be found on page 100.

EASTER BASKETS

The world of "little people" believe that dyed eggs and Easter baskets go hand-in-hand - - - so you'll want to make a special basket for that special wee one in your life.

A fairy's wand waved over a bleach bottle, or your nimble fingers, promise to produce something different and exciting. Miss Darling, the duck, stands 17" high and can hold all kinds of Easter goodies. Turn to page 93 for her instructions.

The orange basket, shown at the right, is a bleach bottle cut off 4" above the bottom. The handle is a ½"x 16½" plastic strip fastened to the basket with brads or staples. The basket is covered with plastic tufts similar to those used on the duck. Tiny flowers and velvet bows add the finishing touch.

FLUTTER-FLIES

You'll enjoy making a collection of these gay, little 4'' to 6'' Lepidoptera. They'll become real home-brighteners when used to accent flower arrangements, or cherished treasures to add to your collection of handcrafted Christmas tree ornaments. It's even fun to transform them into refrigerator magnetics by gluing magnetic tape to the bottom of them. Know-how and patterns can be found on page 110.

BEACH TOTE BAG

Here's an eye-catching, happy accessory to give a touch of originality to your fun-filled hours at the beach.

This chic and charming tote bag, made from plastic bleach bottles, promises to be excitingly useful for carrying and stowing "beachable needables".

Turn to page 108 for instructions and then make this all-time favorite.

103

Don't let this Southern Belle's fragile femininity fool you into thinking she's just another frilly doll. Under the lacy loveliness of her shade shines a light that will brighten your home as much as her sweet appearance will light up your eyes. Because Belle is an outstanding costume doll lamp, instructions are given for her attire as well as for the lamp she so charmingly "dresses up." A doll in the costume of your choice would be just as appropriate, provided it camouflages the lamp base.

Belle's traditional hoop skirt hides a sturdy plastic lamp base which is so simple to make you won't need a "he-man" to help — or hinder your "playing dolls!" A plastic container weighted down with casting plaster can be an ideal lamp base for any hand-me-down, whether it's an old coffee mill, or a childhood doll out of the attic trunk. So salvage a cast-off container. Then turn it into a lamp, and watch it glow over its new lease on life!

"SOUTHERN BELLE" Bedside lamp

MATERIAL NEEDED FOR LAMP

6'' dia. round plastic container with lid
11'' to 12'' doll of your choice
6 ft. electrical cord
12'' electrical tubing, threaded at one end

Socket
Plug
Light bulb
Casting plaster

MATERIAL NEEDED FOR DOLL CLOTHES AND LAMP SHADE

1 yd. scalloped lace, 45'' wide or fabric of your choice
1¼ yds. lining, 45'' wide
8½ yds. lace, 3/8'' wide
3/32'' art foam, 36''x 36''
2 yds. fancy cord
10'' ribbon, 3/8'' wide
All-purpose clear glue

10½'' dia. provincial lamp shade, 6'' high, with harp attached
Cloth flowers of your choice
White string
14 chenille stems, 12'' ea.

DIRECTIONS

1. Place tubing ¾'' from inside edge of lid, and doll's feet slightly in front of tubing. Mark placement for each on lid.
2. Heat one end of tubing over flame and burn hole thru lid for tubing. Cut holes for feet with craft knife. Holes for feet should be wide enough to fit top of doll's knees.
3. Punch hole for cord close to bottom of container. Cover hole with masking tape to prevent plaster from leaking out until it sets. Make a slit large enough for cord.
4. Run cord thru tube. Put tube thru lid. Pull cord out of slit in taped hole. See Fig. A. Pull cord until 3'' remain at top of tubing.

Fig. A Pull the cord out of the taped hole.

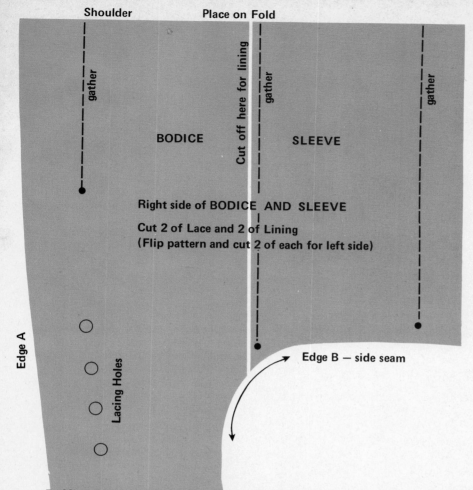

Shoulder   Place on Fold

gather

Cut off here for lining

gather

gather

BODICE   SLEEVE

Right side of BODICE AND SLEEVE

Cut 2 of Lace and 2 of Lining
(Flip pattern and cut 2 of each for left side)

Edge A

Lacing Holes

Edge B — side seam

5. Mix enough plaster to fill container ¾ full. Pour into container. Close lid. Hold tubing straight until it stands without help.

6. Make HAT: Cut a 4½" dia. lace circle. Gather edge to fit crown of doll's head. Secure gathering by threading a needle with end of gathering thread and sewing a knot. Clip thread close to knot. Cut a 17" x 2¾" hat rim. Gather to fit crown of hat. Secure gathering. Sew back seam of hat rim. With right sides together, stitch gathered edge of rim to gathered edge of crown. Turn right side out. Weave chenille stems thru holes in lace along edge of hat rim (optional). Glue chenille ends. Place hat on head. Glue 3/8" ribbon hatband over gathering. Add cloth flowers if desired.

7. Make BODICE: Cut bodice and bodice lining. With right sides of fabric together, stitch bodice lining along edge A. Turn right side out. Gather between ●'s. Adjust fit. Sew side seams. Make eyelet lacing holes on both sides of bodice front. Lace with fancy cord. Place on doll. Pin in back. Lace and tie. Make a small bodice inset to place behind lacing if desired.

8. Mix more plaster and pour into lamp base. Close lid and immediately insert doll's feet thru holes in lid. See Fig. B. Hold doll straight until she stands firm.

9. Make FOAM UNDERSKIRT: Make a double thickness of foam by sewing the 42" x 6" piece onto the 42" x 8" piece as illustrated below. (Illustration 1) Sew a casing along length of single thickness of foam, as illustrated. (Illustration 2) Run string thru casing for waist. Sew back seam, leaving 2" open below waistline.

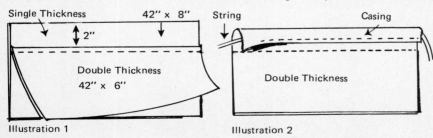

Single Thickness   42" x 8"   String   Casing

2"

Double Thickness
42" x 6"

Double Thickness

Illustration 1   Illustration 2

Place underskirt over doll and tubing. Gather string tightly around waist. Knot and clip.

10. Make SKIRT LINING: Cut two 43" x 10" pieces for two skirt linings. For each piece, edge hem with lace, sew two rows of gathering stitches for waistline across 43" edge, and sew back seam, leaving 2" open at top. Place one lining over underskirt. Gather waist and knot thread. To give skirt shape, whipstitch chenille stems for hoop to wrong side of other skirt lining 1" above hem. Proceed as for other skirt lining.

11. Make LACE SKIRT: Cut a 45" x 10½" piece of scalloped lace. Proceed as for first skirt lining.

12. Attach plug and socket to electrical cord as directed on package instructions. Add bulb.

13. Make SHADE LINING: Cut a 42" x 7" piece of lining. Edge top and bottom 42" lengths with lace trim. Sew gathering stitches 2" below a lace edge. Sew back seam, leaving a small opening where gathering threads meet. Place over shade, with gathering at top. Gather and knot threads.

14. Make LACE SHADE: Cut a 42" x 8" piece scalloped lace. Edge unfinished length with lace trim. Sew gathering stitches 2" below lace trim. Seam as for shade lining. Place over shade lining; gather and knot threads.

15. Wind 1 yd. of fancy cord or ribbon around gathering twice and tie in a bow. Put shade over bulb.

16. Allow lamp base to dry one day before plugging into electrical outlet. □

Fig. B Place doll in plaster-filled base.

CHRISTMAS POINSETTIAS

Cherished memories of by-gone Christmases often include the family fun of making holiday trims and decorations. Treasured hand-crafted projects are proudly displayed and then carefully tucked away for another year.

These 10" to 14" diameter plastic poinsettias promise to be a happy memory-maker, too - - - so save your plastic milk bottles, gather your family together and before you know it, your front door will be the envy of the neighborhood!

MATERIALS
2 one-gal. plastic milk bottles, translucent
2 half-gal. plastic milk bottles, translucent
6 strings of 3 mm strung pearls, 60" per string
3 pcs. 24 ga. floral wire, covered, white
2 foam balls, 1¼" dia.
White craft glue
Epoxy glue

DIRECTIONS
1. FLOWER CENTER: Cut a foam ball in half. Bend a 12" wire into a hair pin shape and push thru the top half round of the ball. See Fig. A. Pull taut. Completely cover the half ball with pearls. To do this, coat the rounded portion with craft glue; start at the center and spiral pearls onto it. See Fig. B. Make three centers.
2. PETALS:
Trace and cut the basic petal patterns of lightweight cardboard. Since the petals of a poinsettia are not alike, the patterns can be altered for variation when you cut the plastic petals. For shape, the petals are cut from the curved surface of the plastic bottles. Cut each petal so the curve is at a little different angle. See Fig. C.

3. Trace a pattern onto the bottle with a soft lead pencil; then cut out the petal with scissors. Cut seven small petals for each of the two small flowers and seven large petals for the large flower. Bend or "pop" some curves in reverse to eliminate the "bottle" look after the petals are cut.

4. Cover your work area with paper; then lightly coat the petals with pearl spray. Allow to dry.

5. Place the petals on a thick piece of cardboard for ease of punching a hole; punch a hole thru the bottom of each petal with an awl or pointed

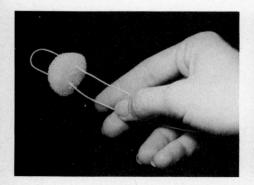

Fig. A   Insert wire thru foam ball.

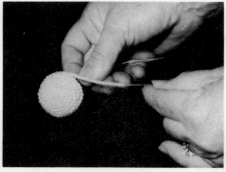

Fig. B   Cover ball with pearls.

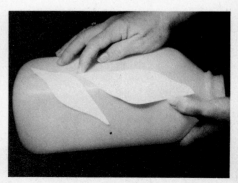

Fig. C   Cut petals from curve on bottle.

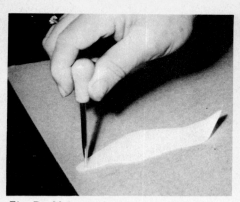
Fig. D Make a hole in the bottom of the petal.

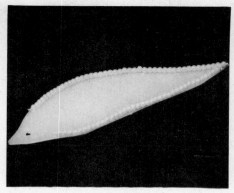
Fig. E Leave the bottom end of the petal untrimmed.

Fig. F Add petals to the wire on the back of the flower center.

POINSETTIA PATTERNS

tool. See Fig. D. The hole will be used for assembly in step 7.

6. Use epoxy glue to trim the outside edge of the petals with strung pearls.* Leave the bottom ¾'' untrimmed as shown in Fig. E. When cutting strung pearls, add a dot of glue to both ends of the cut to keep the pearls from falling off the string.

7. ASSEMBLY: Thread the wire, at the bottom of the beaded flower center, thru the hole of one petal. Draw the petal up snug against the backside of the flower center. See Fig. F. Continue adding six more petals. Tie a knot in the wire to keep the petals from sliding off. Do NOT cut off the excess wire ties.

8. Arrange the petals around the center. Add an application of epoxy between the petals where they touch in the center. If the petals slide around, saturate a little cotton with epoxy and stuff between them. Allow the epoxy to dry.

9. After making two small and one large poinsettias, fasten them to a spray of greenery with the wire ties. Add a red ribbon bow and satin-covered balls to complete the door-piece. □

* It is difficult to glue anything onto this particular kind of plastic. Epoxy glue will hold the pearls in place but handle carefully so they don't "pop" off when the glue is dry. If the pearls come loose, simply re-glue them.

LAZY DAISY TOTE

Make this 12"x 12" tote-all to carry on your next lazy day's trip to the beach!

It is fashioned of white panels cut from bleach bottles, assembled with yarn and decorated with yarn daisies.

This bag will hold your fun-time gear --- swimwear, towels, snacks, and camera.

MATERIALS
5 one-gal. bleach bottles
4-ply yarn, color of
    your choice
½ yd. 36" lining material
Large button

SUPPLIES
1/8" dia. paper punch
Awl or pointed tool
Tapestry needle, No. 18

INSTRUCTIONS

1. Cut off and discard the top and bottom of a bottle. Cut a 4¾" x 19" panel and lay it flat. Repeat for the other bottles.

2. See opposite page and cut master cardboard patterns as instructed. Trace and cut pieces from the plastic panels.

3. With a paper punch or awl, make holes around each piece except the handles. The holes should be about ¼" apart and 3/16" in from the edge.

4. Blanket stitch around each piece with yarn. Put two stitches in each hole and four in each corner hole. See Fig. A.

5. Decorate each of the A and B pieces with a yarn daisy. To do this, use an awl to punch seven holes, ¾" apart, arranged as shown in Fig. B. (The center hole should be larger than the others.) Stitch a lazy daisy with a french knot center. See the diagrams for the lazy daisy stitch and french knot.

6. Whipstitch six A pieces around one B piece for a front panel, stitching on wrong side. See Fig. C. Repeat for back panel.

7. Trim the handles by punching holes 5/8" apart down the center of each strip. Blanket stitch around the edges with yarn. Put four stitches in each hole along each side of the handle. See Fig. D on the next page.

8. ASSEMBLY: Place the three large rectangular pieces end to end and whipstitch them together. Whipstitch a small rectangular piece to each end. This will form the bottom and sides of the bag.

9. Whipstitch the front and back panels into place.

10. Line the bag, as desired.

11. Add a loop for the button to the rounded

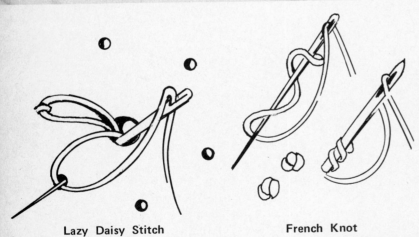

Lazy Daisy Stitch   **French Knot**

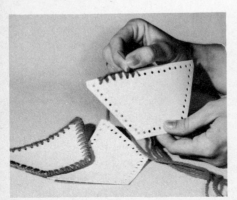

Fig. A Stitch around holes.

Fig. B Punch holes for daisy

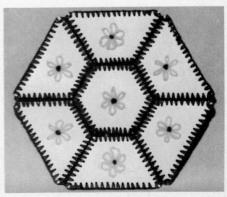

Fig. C Assemble the front panel.

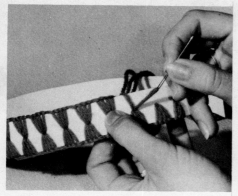

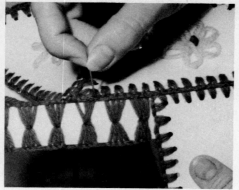

Fig. D Trim the handle with yarn.

Fig. E Sew handle into place.

CUTTING INSTRUCTIONS

A-PIECE, cut 12
B-PIECE, cut 2
CLASP, Cut 1

Also Need:

HANDLE, 1" x 19"; cut 2
BOTTOM & SIDES, 3¼" x 5¾"; cut 3
TOP SIDES, 3¼" x 4"; Cut 2

end of the clasp and whipstitch the clasp into place.

12. Whipstitch handles into place. One end will be secured to the top of the small rectangular piece and then along the exposed edge of the front panel. See Fig. E.

13. Sew a button onto the front.   □

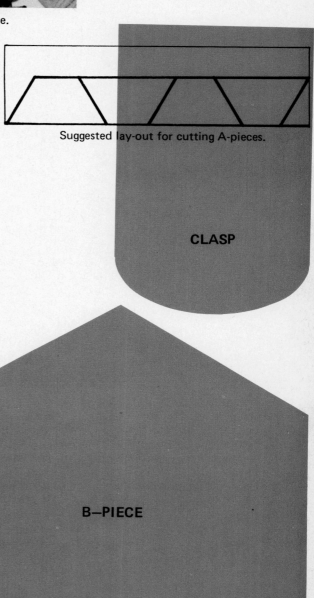

Suggested lay-out for cutting A-pieces.

CLASP

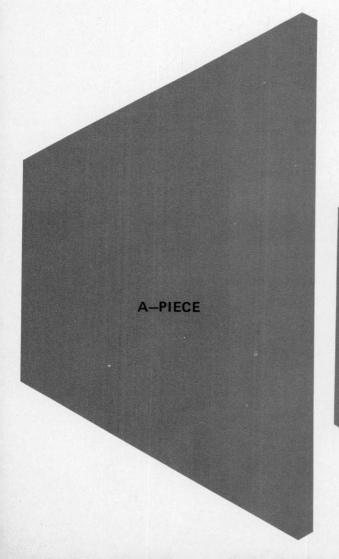

A—PIECE

B—PIECE

FLUTTER-FLIES

Ever have the flutter-flies? You should! They'll get rid of any uneasy sensation you may have over finding interesting decorating novelties for your home or office. Just make these bejewelled beauties out of ordinary plastic milk bottles. Then trim their wings with shimmering metallic cord and sparkling beads. Add a touch of jewel-toned wax gilt, and Presto! You have "the flutter-flies" like those shown on this page.

BUTTERFLY MATERIALS
1 gal. plastic milk bottle
Epoxy glue
2 jewels
1 lg. chenille bump, black
1½ yds. tinsel-tex braid, gold
Optional: Rub-on metallic wax,
    color of your choice

DIRECTIONS FOR THE BUTTERFLY
1. Cut a full-size pattern of heavy paper. Place the pattern lengthwise on the curve of the plastic bottle. See Fig. A. This will give shape to the wings. Trace and cut one.
2. OPTIONAL: Color the wings with rub-on metallic wax.
3. Glue gold braid around the edge of the wings and as a design onto the wings. Glue a jewel onto each wing.
4. Cut two 4½'' lengths of braid for antennae. Shape one end of each into a curl and glue the other ends to the top of the body.
5. Glue the chenille bump between the wings as a fluffy body. ☐

Place on fold

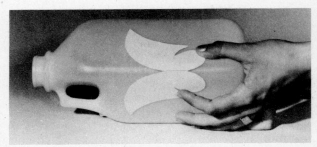

Fig. A   Place the pattern on the curve of the bottle.

DRAGONFLY MATERIALS
1 gal. plastic milk bottle
While craft glue
3 Chenille stems, green
Rub-on metallic wax; blue,
    green and silver
Wooden beads:
    1- 18mm, green
    5- 12 mm, green
    4- 10 mm, green
    2- 8 mm, green
    3- 6 mm, green
    2- 6mm, blue, eyes

DIRECTIONS FOR THE DRAGONFLY
1. Repeat step 1 as for the butterfly. Exception, cut two sets of wings.
2. Color one set of wings with blue metallic rub-on and the other set with green. Shade both sets with silver.
3. Form the body by threading the green wooden beads, in graduated order, onto a 6½'' length of chenille. Allow ¾'' to extend from the mouth of the large bead head. Glue the end beads in place.
4. Cut a 1'' piece of chenille and glue into the mouth making a forked tongue effect. Glue the two blue bead eyes onto the head.
5. Cut three 5'' lengths of chenille for legs.
6. To assemble, place one set of wings on top of the other allowing the bottom set to extend ¼'' beyond the top set.
7. Place the bead body on the wings. With an awl, pierce a hole thru the wings between the second and third bead. Push one end of a leg thru the hole; loop the other end over the bead body and down thru the same hole. See Fig. B. Pull taut. This makes two legs. Repeat for the other legs placing one set of legs between the third and fourth bead and the other between the fourth and fifth bead. Bend the legs into shape.
8. Touch up the body with silver metallic rub-on. ☐

Place on fold

chenille loop

Fig. B   Loop chenille legs over the bead body.

Plastic Bottle Fun

The projects shown on the following pages were selected for your pleasure. Some are "quickies" and some not so quick. We hope these ideas will stimulate your own imagination as you have fun with empty bleach bottles, soap or detergent bottles, and just plain plastic bottles.

Plastic bottles of many sizes and shapes are available to anyone---as they are usually discarded. Wash them, soak off labels, and clean off the glue before starting to work with them. The commercial glue may be cleaned off with acetone or lacquer thinner.

Use a sharp knife or razor blade to cut the plastic. Warm the bottle in the sun or fill with hot water before cutting it. You will find that it cuts more easily when warm. A small coping saw is very useful when cutting off the handles or the screw top.

Bottles may be painted with any good craft paint. Their smooth surface makes pasting and gluing fun and easy. Jewels, Gold Paper Lace, felt, rick rack, scraps of paper or material, artificial flowers, feathers, or foil---any or all of these materials combine well with plastic. With a little imagination and patience you can have a world of fun.

Materials Needed:

Asstd. Plastic Bottles
Knife
Scissors
Stapler
Craft Paint
Paint Brushes
Asstd. White Foam
White Glue
Masking Tape
Felt, Sequins, Gold Paper
    Lace, Braid, etc.

Barnyard Scene

To make the barn, cut a large opening in the bottom half of a gallon bleach bottle. Cut doors in the top half for a hay loft. Make hinges out of black paper. Cut a round piece of White Foam for a floor for the hay loft. Glue to hold. Cover floor with excelsior for hay.

Decorate edges of doors and inside of barn with paper that looks like wood. A self-adhesive product is available at many stores. Make a cupola out of a little square box. Add a peaked roof. Cover with wood-grained paper. Decorate with a weathervane, birds, etc. cut out from cards or magazines.

Cut a free form White Foam base and glue barn to it. Cover base with glue and gravel or paper that looks like rocks. Make a little well and outdoor toilet of cardboard boxes covered with wood-grained paper. Toy fences and barnyard animals may be purchased at many hobby and toy stores. Use your imagination to complete the scene.

The scene in the picture to the right has a wood pile to one side, a farmer and his wife just coming out of the barn, and tiny birds roosting on a branch above the door.

Curler Bag

MATERIALS NEEDED:

1 gallon bleach bottle
18" x 18" piece of material - cotton preferred
1 plastic curler
1-1/2 yds. cord, braid, or ribbon - for drawstring
Red and black felt for mouth and eyes
White glue
9" piece of chenille - hair
Rick-rack for trim

DIRECTIONS:

1. Cut top off bleach bottle, discarding the neck and handle. Fig. 1. Note that one side is left higher than the other. The taller side is the front.

2. Sew two edges of the material together making a tube that will fit snugly over the bottle.

3. Now cut a hole out of the material so that you have a place for the face. Fig. 2. Glue edges of the material to the bottle. Glue rick-rack around face.

4. Make a 2" hem at the top of the material. Stitch a row half way down the hem so that you have a tunnel through which to run the drawstring. Slip drawstring through tunnel and tie ends together. Fig. 3.

5. Cut eyes and mouth out of felt and glue to face.

6. Wrap curler with chenille and sew to front of cap. Make a small bow of material and sew above curler.

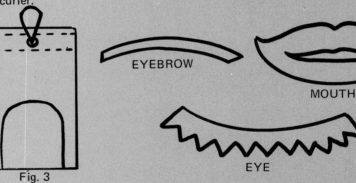

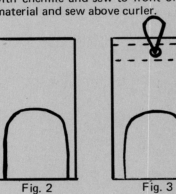

Fig. 1   Fig. 2   Fig. 3

EYEBROW

MOUTH

EYE

Covers for Hair Spray, Cleansers, etc.

HAIR SPRAY COVER

Select a plastic bottle in which your can of hair spray will fit. Cut off the top. Decorate with base metal, braid, beads, jewels, etc. as desired.

CLEANSER CAN COVER

Choose a plastic bottle that is large enough to fit around a can of cleanser. Cut off the top. Glue a band of gold braid near top and bottom. Glue a spray of sea shells and coral to one side. Add tiny flowers made of sea shells. See suggested flower formations below. Now spray with pearlized paint. Glue pearls and jewels here and there to add a bit of glamour.

Fancy Fido

MATERIALS NEEDED:

Quart size plastic bottle
6" x 6" brown felt - ears and nose
2" x 2-1/2" red felt for tongue
12" velvet tubing for tail
Chenille stem for tail
2 moving eyes - 3/4" dia.
4 plastic bottle caps or corks - feet
Pre-cut felt flowers and circles
2" x 4" black paper or felt - eyes
Sequins, pins, and white glue

DIRECTIONS:

1. Cover screw cap "nose" with felt and add 2 sequin nostrils.

2. For tail, bend chenille stem over 1/4" at end and run through velvet tubing. Sew end closed. Pull tubing tight over chenille and secure at other end. Make hole in plastic bottle for tail and glue into place. Bend to shape.

3. See patterns and cut ears and tongue out of felt. Glue tongue into place. Cut 3/4" slit for each ear lengthwise on the bottle about 1-3/4" apart. See Fig. 1. Insert the ears and glue to secure.

4. Glue eyes and black circles into place. Glue bottle cap or cork legs into place. Decorate with felt flowers, circles, and leaves.

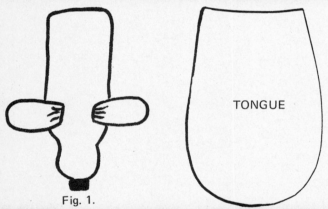

Fig. 1.

TONGUE

EARS
CUT 2

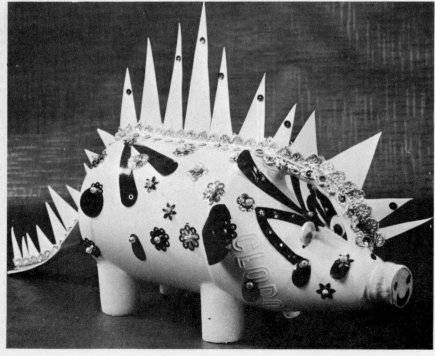

Dinosaur

The dinosaur is made very much like the dog. The eys are inserted and glued into slits so that they stand upright. A tail and backbone with spines are cut out of another plastic bottle. These are fastened to the center of the back with masking tape.

The masking tape is then covered with Gold Paper Lace and sequins. Eyebrows and ears are cut out of black paper or felt and glued to head. He may be decorated any way desired. Your imagination can be as fantastic as you wish.

Mr. Brown Bangs

2 bleach bottles - 1/2 gal.
5 White Foam poles---
   1" x 6" for trunk & legs
2" x 8" pink cardboard
1 white chenille stem
4-1/2" x 19" red felt
2 moving eyes (3/8")
10" brown ribbon - bangs
Black construction paper
White glue
Trim: Gold Paper Lace,
   flowers, ribbon bows,
   and sequins.

DIRECTIONS:

1. Cut handle and screw cap of the bottle off at an angle. Using pattern, cut the trunk out of the White Foam. Round edges and smooth with sand paper. Taper so that end of trunk is about 1/2" wide. Glue into hole where cap was cut off.

2. Cut ears out of the other plastic bottle. Make two slits in the head of the first bottle about 2" on either side of the seam line. Glue ears into slits.

3. Cut the brown ribbon into 5 pieces. Cut one end of the ribbon into notches and glue to forehead. The ends should all meet at the top. Glue a bow where ribbon ends meet.

4. Cut black construction paper eyes. Glue to the head about 1/2" from the trunk and 1/2" from the ear. Glue moving eyes to the center of the black paper. Add a few flowers to the trunk where it connects to the head and add Gold Paper Lace to each ear for earrings.

5. Take a hot ice pick and make a hole in the bottom of the bottle even with the seam line. Glue chenille stem in place for the tail. Add a bow to tail if desired.

6. Legs: Shape White Foam pole legs to fit the contour of the body and glue to body. Cut pink cardboard feet and glue to bottom of legs.

7. Blanket: Fringe the ends of the red felt in approximately 3" slits all along the edge. Place felt on the back of the elephant and glue to secure. Glue gold paper lace medallions and edging to blanket for trim.

NOTE: You can change the expression of your elephant by the slant of his ears and the angle in which his trunk is attached.

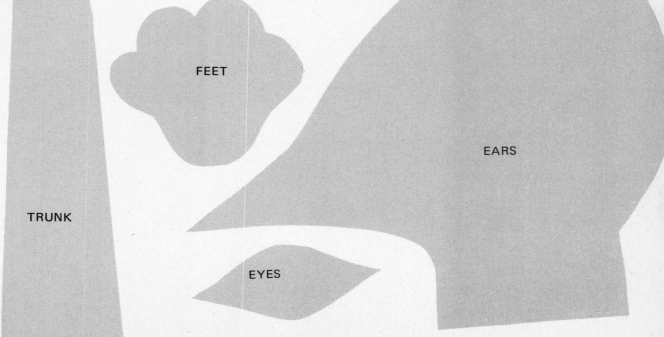

TRUNK

FEET

EYES

EARS

Bird Cage

MATERIALS NEEDED:

- 1/2 gallon plastic bottle
- Assorted small plastic flowers, leaves, & butterflies
- 1/4" x 8" doweling
- Gold Paper Lace trim
- 1 bird
- Sequins
- White glue
- Scrap White Foam

DIRECTIONS:

1. Cut 3 holes in bottle, 2-3/4" diameter each. One hole in center front and one on either side---equal distance apart.

2. Glue Gold Paper Lace around bottle at indentation lines above & below holes. Add other trim as desired.

3. Run doweling through bottle and put bird on his perch.

4. Decorate with flowers, leaves, butterflies, etc.

5. Insert piece of scrap foam into top neck of bottle and anchor flowers in it. (Floral clay may be used to help hold flowers.)

The bottle at the left is turquoise, the flowers are pink with a pink & yellow butterfly. The yellow bird sits on a pink perch made from a small dowel stick.

To make a flower basket, cut top off of bleach bottle leaving about 6" at the bottom. Cut slits 3/4" wide from top to within 1" of bottom. Fig. 1. Wrap strips of plastic around a pencil so that they curve down. Fig. 2. Continue until all strips have been curved.

Make handle by cutting a strip of plastic 1" wide and 18" long. Place inside basket and staple or glue to hold.

Fig. 1

Fig. 2

Toaster Cover

Make the doll of your choice to resemble a teenager with a pony tail or pig tails---or even a colonial lady simply by changing the face mask and costume design.

MATERIALS NEEDED:

1 plastic detergent bottle
3" White Foam egg or ball
1 plastic face mask
Heavy cardboard - 4" x 5"
36" covered wire - 18 ga.
1 yd. percale - dress
6" x 18" organdy - apron
  and kerchief
Crepe paper or tissue paper
  for hands
Trim: rick rack, loop earrings and cameo pin

DIRECTIONS:

1. Cut off top 4" or 5" of plastic detergent bottle for doll bodice. Fig. 1.
2. Force White Foam egg down over neck of bottle for head. Fig. 2.
3. Place face mask on the front of the egg and pin to hold.
4. Make two small holes in both sides of the bottle near the shoulders.
5. Fold wire in half and push both ends through the holes in one side and through the bottle and out the other side. Fig. 3. Twist wires together so that they stay in place for arms.
6. Wrap ends of wire arms with crepe paper or tissue paper and secure with glue for hands. Fig. 4. You may cover with an old nylon stocking for a nice flesh tone.
7. Cut a piece of heavy cardboard about 4" x 6". Cover this with material.
8. Make holes around the bottom of the doll bodice with a needle that has been heated over a flame.
9. Sew doll bodice to center of cardboard pad. Fig. 5.
10. For blouse, cut a piece of percale 5" x 8". Cut a small hole in the center for the neck. Slit down the back. Cut two sleeves 4" wide and 5" long. Sew to both sides of the blouse. See Fig. 6. Sew both side and underarm seams. Place on doll gathering sleeves around arms. Sew or pin back together overlapping material.
11. Sew two pieces of percale 12" x 30" together at both sides for a skirt. Gather at top to fit waist of doll. Hem skirt and trim with rick rack.
12. Cut a piece of organdy 6" x 10" for the apron. Hem along three sides. Gather along top edge to fit a waist band 3" in length. Now cut two pieces of organdy 1" x 12" for apron ties. Fold pieces in half lengthwise and stitch together so that ties are 1/2" x 12". Sew to sides of waist band. Tie apron around waist of doll.
13. Make a kerchief out of a 6" square of organdy. Fold in half to form a triangle and place around neck. Secure with cameo pin. Make a bandana out of percale the same as the kerchief and tie around head. Add loop earrings.

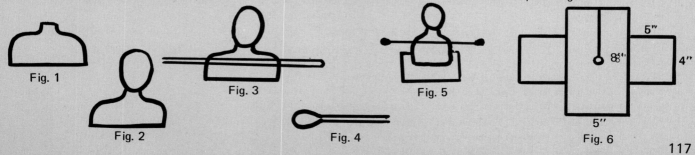

Fig. 1   Fig. 2   Fig. 3   Fig. 5   5"   8" 4"

Fig. 4   5"   4"

Fig. 6

Crayon Clown

This Crayon Clown makes a perfect gift for the younger set.

MATERIALS NEEDED:

    1 plastic detergent bottle
    10" percale yardage - 36" wide
    2 pieces felt - 9" x 12" ea. - feet, hands, hat
    3" x 72" net - ruffle
    15" piece 18 ga. wire - arms
    1 plastic clown puppet head
    Cardboard - feet
    Crayons, pencil, eraser, pad
    White glue, needle and thread

DIRECTIONS:

1. Fill bottle with plaster or gravel to give clown weight. Place puppet head over neck of bottle and glue to secure.

2. Cut feet out of cardboard and felt and glue together. Glue these to bottom of bottle.

3. Punch a hole in both sides of the bottle at the shoulders and push the heavy wire through the holes. Let the two ends extend for arms. Secure with glue or masking tape.

4. Cut 2 pieces of percale making the front and back to the robe. See pattern. Sew shoulder seams. Now sew in the sleeves. Sew side seams and sleeve seams.

5. Sew a 3-1/2" x 24" strip of felt around the bottom of the robe. Sew only at the bottom. Make pockets in this felt strip to fit pencil, eraser, and pad. Sew a seam from top to bottom. After these pockets have been made, make slots 3/4" wide all around the rest of the way for the crayons.

6. Put the robe on the clown. Gather at the neck and arms to fit. The robe and sleeves will fit loosely.

7. Cut hands out of felt and sew edges together. Pad with a bit of cotton or paper and slip over ends of wire arms.

8. Take a strip of 3" wide and 36" long net and gather down the middle for a collar. Place collar around the neck. Make two ruffles out of 1-1/2" x 12" net for arms.

9. Cut hat out of felt. Lap straight edge of hat over dotted line to make a pointed hat. Sew or glue and secure to clown's head.

PLACE ON FOLD

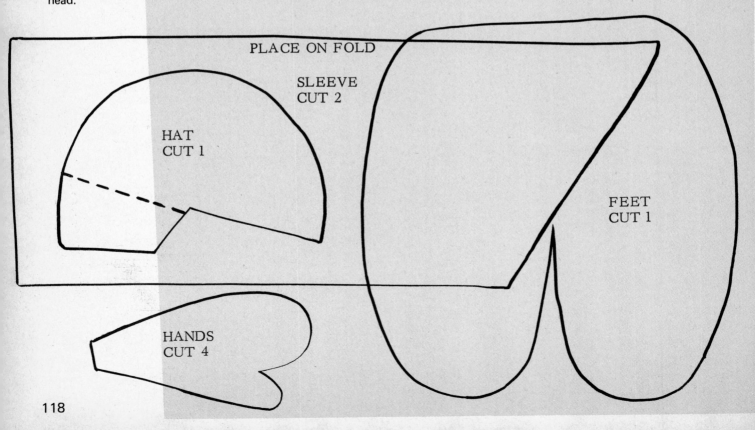

PLACE ON FOLD

SLEEVE
CUT 2

HAT
CUT 1

FEET
CUT 1

HANDS
CUT 4

Baby Shower Bassinet

Fill bassinet with real flowers or use a baby doll as shown in the photo of the finished centerpiece.

MATERIALS NEEDED:

1 large bleach bottle
1" x 12" x 16" White Foam
1 yard nylon net - 72" wide
5 yards ribbon (5/8")
10 yards narrow ribbon
2 pieces 18 ga. white wire-
   30" long
White Glue
1 bunch velvet forget-me-nots

DIRECTIONS:

1. Cut bleach bottle in half lengthwise, discarding the part that has the handle.
2. Glue this bottle to White Foam base with the open side up. This keeps it from rocking from side to side. If you desire, you may cut the corners of the base the same as the bottle.
3. To make a frame for the hood of the bassinet, curve both pieces of wire into a "U" shape. Force the ends into the White Foam base. The first wire should be at the back of the bottle and the second wire should be 3" from the first. Put a dab of glue where they enter the base. Either sew or wire to the bassinet near the top to make them stay in place.
4. Now cover the raw edge of bottle by gluing a ribbon over the edge. Cover wires with ribbon. Glue a piece to the top and one to the bottom so that the wire is completely covered.
5. To cover the hood, cut a piece of net 12" x 72". Fold it in half so you have a double piece 12" x 36". Gather along one edge with a basting thread. Pin gathered edge to front of wire hood and sew to ribbon.
6. Fold net down over the back of the hood. Pin and sew through net to the ribbon edge of bassinet.
7. Cut three pieces of net 6" wide and 72" long. Gather near one edge making a ruffle to go all around the bassinet. Sew ruffle to ribbon at top edge of bassinet.
8. Pin ribbon around edge of White Foam base.
9. Make three bows out of the narrow ribbon. Sew to each side of hood and one to front of bassinet. Add little clusters of forget-me-nots to the bows. Glue single flowers all along the top edge of bassinet and up over edge of hood.

Pin Cushion Chair

This Pin Cushion Chair is perfect to use in your bath or boudoir.

MATERIALS NEEDED:

Quart size bleach bottle
3" square White Foam - 1-1/2" thick
Trim: net, rick rack, sequins,
   gold paper lace

4" square Art Foam or felt
White glue
Pins

DIRECTIONS:

1. Cut bottom of bottle in a chair shape. The chair back should be about 6" high and 5" wide. Cut a circle of White Foam to fit inside the chair bottom. It should be about 3" in diameter. Cover the top of the White Foam with a piece of felt or Art Foam. Bring edges down over the White Foam and pin to hold. Force this down inside the chair bottom.
2. Pin and glue a ruffle of net around the base of the chair. Trim with rick rack and sequins. Gold Paper Lace daisy strips may be glued all around edge of the chair back if desired.

Head Hunter Mask

To use the Head Hunter Mask as a string holder, punch a hole in the back to the tongue to allow string to be dispensed.

MATERIALS NEEDED:

Gallon size plastic bottle
2 smaller plastic bottles with handles
Gold Paper Lace - trim
1" square red felt - tongue
2 moving eyes
9" White Foam disc
1 chenille stem - hanger
White glue
Pins
Sequins
Base metal - ears

DIRECTIONS:

1. Take disc of White Foam and cut out 6 notches as in Fig. 1. This leaves pieces about 2-1/2" wide and 1-1/2" deep extending out from the center circle. Fig. 2.

2. Cut off top of gallon size bottle. Fig. 3. This makes the face with the handle for the forehead and the bottle opening for the nose.

3. For eyes and horns, cut the handles and bottle necks off of two smaller bottles. Fig. 4. Make two round holes the same size of the bottle necks on each side of the face. Fig. 5 and 6. Force the bottle necks into these holes and secure with glue. This makes two horns on each side of the face. Fig. 7.

4. Cut a piece of scrap White Foam to fit inside the eyes. Now glue the two moving eyes to the top of this. Cover nose by gluing a piece of cardboard or white paper over it. Glue two sequins to the nose for nostrils.

5. Cut two strips of plastic from a bottle. Make one 3-3/4" x 3/4" and the other 4-1/4" x 3/4". Glue or staple these together as in Fig. 8. When dry, glue to face just below the nose. Fig. 9. Glue a little felt tongue inside the mouth.

6. Place face on top of the White Foam disc that you cut at first. Place the notches so that one is on top and one on the bottom and two on each side. Glue and pin to hold in place.

7. To make the pointed ears, cut 5 pieces of plastic as in Fig. 10. Take the two 1" pieces and glue or staple together. This makes the ear curve and makes the point stand up away from the bottom. Fig. 11.

8. Place these next to the face and pin to the White Foam projections as in picture. Decorate mask with Gold Paper Lace and sequins. Cut two ears out of paper or base metal as in picture and glue to face. To make hanger, take a 5" piece of chenille stem and force both ends into the White Foam.

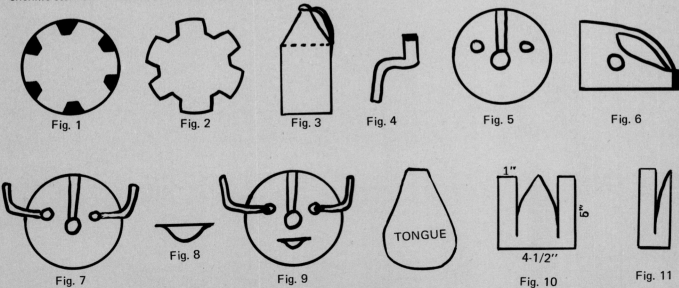

Fig. 1 Fig. 2 Fig. 3 Fig. 4 Fig. 5 Fig. 6 Fig. 7 Fig. 8 Fig. 9 TONGUE 1" 4-1/2" Fig. 10 Fig. 11

Indian Drum

MATERIALS NEEDED:

2 large plastic bleach bottles
2 wooden dowel sticks - 1/2" x 10"
Leather thong, plastic lacing, or twine
2 large wooden beads
String, wire, or chenille stem

Leather or old inner tube
Bright colors paint - poster or enamel
Feathers
Cloth, felt, or Art Foam

DIRECTIONS:

1. Cut top off one bleach bottle leaving bottom between 6-1/2" to 7" high. Take second bottle and cut off bottom 1-1/2". Force this shallow piece over open end of first bottle so that you have a drum with both ends covered.

2. Paint Indian design on sides of drum.

3. Cut 2 circles of leather or old inner tube about 1/2" larger than top of drum. Make holes with punch or ice pick around edge of both pieces. Be sure that they match up and that you have the same number of holes in both pieces.

4. Place drum in between both circles. Loosely weave, lacing up and down through holes. Tighten lacing gradually.

5. Beaters: Fasten wooden bead to one end of dowel stick. Use a small tack or glue to secure. Place a square of cloth or Art Foam over bead and secure ends by wrapping chenille stem or wire around stick.

Indian Noisemaker

Indian noisemaker is made from a small detergent bottle. The neck of the bottle is cut off so the top will be flat. Slip a dowel stick through slits cut in both ends of the bottle. Paint designs as desired. Drop pebbles or beans into rattle through hole in one end and put a bit of tape over hole. Decorate with yarn, raffia, feathers, etc.

3 Wise Men

MATERIALS NEEDED:

3 detergent bottles - approx. 10" high
3 White Foam balls - 3" dia.
3 pieces felt - 9" x 12"
White glue
Base metal
Black construction paper

Angel hair
3 chenille stems
Trim: Glitter, jewels,
    Gold Paper Lace
Pins

DIRECTIONS:

1. Glue the White Foam balls to the necks of the detergent bottles. Decorate front of bottles with glitter, jewels, and Gold Paper Lace if desired.

2. The robes are made out of 9" x 12" pieces of felt gathered around the neck about 3" down from the top. Wrap felt around back of bottle and tie at the neck with chenille stem. Fig. 1. For the first Wiseman, fold tips down and round corners of collar. For the second Wiseman, fold collar all the way down. For the third Wiseman, fold the collar all the way down and cut a slit in the back and round corners.

3. Cut eyes out of black construction paper and glue to face. Glue some angel hair to head for hair.

4. The headpiece for the first Wiseman is just a band of base metal decorated with jewels and Gold Paper Lace. The headpiece for the second Wiseman consists of a piece of felt pinned to the head and falling down over the shoulders with a base metal crown. The crown has four strips of base metal stapled or glued to the edges. Fig. 2. The other ends are pinned to the top of the head making four loops standing up. Finish by putting an Xmas ball right in the center. Decorate with Gold Paper Lace and jewels. The third headpiece is cut as in Fig. 3 with both ends stapled together to make a circle. Decorate with jewels.

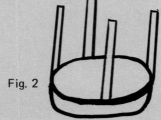

Fig. 1   Fig. 2

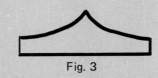

Fig. 3

Television Light
or
Bathroom Night Light

MATERIALS NEEDED:

1 large bleach bottle
3" square White Foam - 1" thick
8" square White Foam - 1/2" thick
White glue
Parchment, self-adhesive paper or
   acetate for windows
Narrow Gold Paper Lace
Gold Paper Lace - 1" wide
Gold Paper Lace Medallions
Sequins and jewels for trim
1 small Xmas ball for top
1 white chenille stem
1 small light bulb

DIRECTIONS:

1. Cut bottom off of bleach bottle. Place this on top of 1/2" thick White Foam square. Draw around this and cut out the inside. Fig. 1. Smooth off any rough edges with sand paper or scrap White Foam. Be sure that the bottle fits in this hole.

2. Place this piece with the hole in it, on top of the 1" thick square of White Foam. Fasten with glue and short pieces of chenille stem. Fig. 2.

3. Decorate sides of base with wide strip of Gold Paper Lace and sequins or jewels as desired. Glue a Gold Paper medallion to each corner.

4. Place the socket end of the cord in the center of the base. Fasten with "U" shaped pieces of chenille stem. Screw small light bulb into socket. Fig. 3.

5. Cut windows out of parchment or self-adhesive paper and glue to sides of bottle as in the picture. Decorate edges with narrow Gold Paper Lace. Leading and stained glass paint may also be used.

6. Remove screw top from bottle and cut a piece of White Foam to fit inside the hole. Glue in place. Force the stem of a small Christmas ball into the center of the White Foam. Use a dab of glue to hold. Decorate top of bottle with Gold Paper Lace, sequins, and jewels.

7. Place the decorated bottle over the light so that it fits into the depression in the White Foam. You will find that the cord is in the way --- so, cut a small notch at the bottom of the bottle so that it will fit over it. Place the handle to the back of the light. Fig. 4.

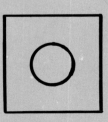

Fig. 1

Fig. 2

Fig. 3

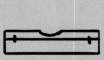

Fig. 4

HP-171

123

Bells and Bows

Cut opening in gallon size bleach bottle as in the picture. Cut 1" thick piece of White Foam to fit bottom of bottle and glue inside. Decorate arch with Gold Paper Lace border. Cut stained glass window out of an old greeting card or use a gold medallion and glue to back of arch.

Stand ceramic or plastic bride and groom to the back of the bottle. Add bows, bells, and tiny artificial flowers for decorations.

Cut off the neck of the bottle and glue a gold Xmas ball over the opening. Glue stained glass paper windows to each side.

Cut posts out of paper drinking straws and make a row of 3 down each side of the aisle. Glue a bead or small Xmas ornament to the top of each post. A small chain or cord may be fastened to each post. Outside of bottle may be decorated as desired.

Church Bank

MATERIALS NEEDED:

1 large bleach bottle	Small bell
Old greeting cards - windows	12 pearl beads
Gold Paper Lace	26 gauge wire - cross
2-1/2" sq. cardboard box	White Glue

Fig. 1

DIRECTIONS:

1. Cut windows and doors for church out of old greeting cards. Glue to sides of bottle. Have the handle to the back of the church. Make frame around windows and doors with Gold Paper Lace.

2. Cut a slit near the top of bottle for money. Outline with Gold Paper Lace. Note: To remove money from bank, cut a hole in the bottom. Seal with tape.

3. For bell tower, take a small cardboard box and cut windows in all 4 sides. Cut 2 sides to a point so that you will have something to help hold up the roof. Cover with wrapping paper or paint. For the pointed roof, cut cardboard to fit over top of box. It should be about 3-1/2" wide and long enough to fold in half and still have about 1/2" left to hang over. Cover with paper or paint. Slip down over tower and glue to hold. Glue tower to top of screw top of bottle.

4. Cross: Take a piece of wire and string one bead in the center. Twist wires together tightly. Fig. 1. String 7 more beads onto the wire. Make cross piece just the same and wire to the first piece. Fasten cross to top of bell tower by spreading wires apart and sticking them down through the top of the roof. Twist ends of wire together and hang bell to wire ends.

124

White Church

MATERIALS NEEDED:

- 2 half gallon bleach bottles
- 5 pictures of church windows --
  (from old greeting cards)
- 10" x 15" heavy cardboard
- 2 pieces White Foam - 12" x 12" x 1/4"
- 3 pieces White Foam - 12" x 12" x 1/2"
- 10" White Foam pole - 1" dia. - posts
- 2" White Foam block
- Small bell
- 12 pearl beads
- 26 gauge wire for cross
- White chenille stems
- White glue
- Small flowers, plants, & pebbles

DIRECTIONS:

1. Cut handles off of both bottles. Cut a 3" strip out of one side of both bottles. Fig. 1. Cut a door out of one just opposite from opening. The door is 3" x 4". Fig. 2. Cut door piece in half lengthwise so that it will open. Fasten back in the hole. Make hinges from small pieces of tape or cloth. Fig. 3.

2. Cut a piece of 1/2" White Foam 7" x 12". Glue to center of 10" x 15" cardboard. Fig. 4. Paint, glitter, or cover with paper that looks like rocks or brick.

3. Place both bottles on top of the base. Put open parts together and glue or staple to hold. Fig. 5.

4. Cut a piece of White Foam 7" x 9-1/2". Cut holes out of this so that it will fit down over the two bottles for a ceiling. Push it down so that it is 5" from the base. Hold up with 5" posts cut from White Foam pole placed at corners. Fig. 6.

5. Cut 2 triangular pieces 5" x 6" out of the 1/2" White Foam. Fig. 7. This is to hold up the roof. Fig. 8.

6. Roof: Take 2 pieces of the 1/4" White Foam. Place on top of ceiling so that they come together to make a peak at the top. Glue together. Use pins to hold till dry.

7. Bell tower: Use the 2" block and cut a "V" out of it to fit the roof. Fig. 9. Glue to hold. To make the windows and peak of tower, cut two 1/2" thick pieces of White Foam to fit top of block. Cut 2 windows out of it and fasten to front and back of block with glue and short pieces of white chenille stem. Fig. 10.

8. Make a cross out of beads and wire and fasten to top of tower. Hang bell inside. Glue windows (cut out of old greeting cards) to sides of church. Decorate roof, windows, and doors with Gold Paper Lace. Glue plants, flowers, and small stones to base of church.

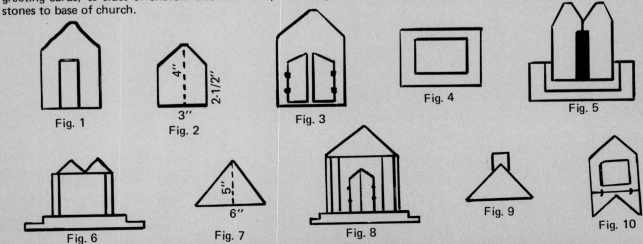

Fig. 1

4"
3"
2-1/2"
Fig. 2

Fig. 3

Fig. 4

Fig. 5

Fig. 6

5"
6"
Fig. 7

Fig. 8

Fig. 9

Fig. 10

Silly Piggy Bank

MATERIALS NEEDED:

1/2 gal. plastic bottle with handle removed
1-1/2" dia. flat bottle cap for hat
4" x 6" red felt for mouth, blanket & feet
4" x 6" green or blue felt for hat & blanket
4" x 6" brown felt for ears & eyelashes
1-1/2" x 1-1/2" pink felt for nose
Pre-cut asstd. felt flowers & leaves
2 moving eyes (3/4" dia.)
1" x 6" White Foam pole for legs
1 bright colored chenille stem for tail
8 gold beads for decor on feet
12" light brown ribbon for bangs
Gold Paper Lace
White glue, and pins

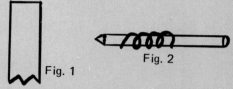

Fig. 1

Fig. 2

DIRECTIONS:

1. Remove bottle handle. Cut money slot in center back if pig is to be used as a bank.

2. Cut hat crown, brim, and band out of green or blue felt. Glue crown to top of bottle cap. Glue band around edge. Place crown on hat brim and glue.

3. Cut four 1-1/2" pieces of 1" diameter White Foam pole for legs. Use serrated knife and sand paper and shape one end of each leg to fit contour of bottle. Glue in place.

4. Decorate feet with 1/2" bands of red felt.

5. Cover screw cap of plastic bottle with pink felt and add 2 sequins for nostrils. Glue Gold Paper Lace trim around neck of bottle (see photo). This will cover hole left after removing bottle handle. Cut and glue red felt mouth into place. Glue eyes in place.

6. Using pattern, cut ears and eyelashes of brown felt and glue into place. Ears may be at any angle desired. Some like them high and some like them low.

7. For decorative blanket, use pinking shears and cut 1-1/2" x 1-1/2" piece of blue or green felt and a 2-1/2" x 3-1/2" piece of red felt. For bank, cut coin slot in center of each piece of felt. Glue red felt on top of pig's back. Glue smaller piece of felt on top of this. Decorate with sequins and beads.

8. Cut five 2-1/4" lengths of brown ribbon. Trim one end of each as in Fig. 1 for bangs. Glue into place on top of head and add hat.

9. Glue decorative pre-cut flowers and leaves to pig as desired. Add extra color and glamour by gluing sequins to each flower.

10. For curly tail, wrap chenille stem around a pencil and then remove. See Fig. 2. Make a hole in the plastic bottle for a tail and glue in place.

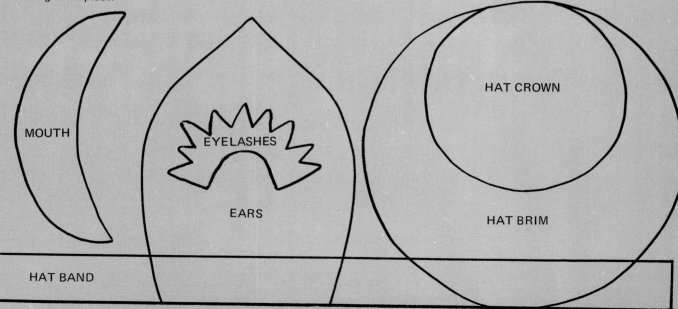

MOUTH

EYELASHES

EARS

HAT CROWN

HAT BRIM

HAT BAND

Children's Party Table Centerpiece

Place on fold

MATERIALS NEEDED:

1 gallon plastic bottle
1" x 6" White Foam disc for base
2" x 11" White Foam pole for center
8 plastic drinking straws
6" x 12" gold deco-foil
1 sht. red velour paper
8mm gold sequins
5mm gold sequins
3 dz. large fancy red sequins
White glue and pins
Horses or animals (cut from cards, etc.)

DIRECTIONS:

1. Cut bottom off bottle leaving 1" rim; this will be the carousel base. Cut top off bottle at indentation line leaving on handle and cap. This will be carousel roof.

2. Cut velour paper 1/4" wide and wrap spirally around center pole, securing with pins or glue. Small, 5mm sequins may be pinned to this if desired.

3. Cut 2" diameter circle in center of White Foam base and glue center pole into place. Glue foam base in bottle bottom and secure with pins from the underside.

4. To cover base, cut velour paper circle 1/2" larger than base. Every 1/2" make a 1/2" cut for the turn-down tab around outside edge. Cut 2" diameter circle in middle. Lower velour paper down over pole onto base. Turn tabs over edge and pin through bottle into base.

5. See scallop pattern and cut strip of velour paper scallops 21" long, to go around outside edge of base. Pin into place through large 8mm gold sequins as in photo.

6. Glue or pin 5mm sequins around base of center pole.

7. See diamond pattern and cut 20 from deco-foil. Staple around the carousel top, short points down. Glue large fancy red sequins over staples.

8. Cut six 1/2" triangles and six arrow shapes. Arrange and glue around top of bottle and cap. Glue fancy red sequins in between as desired.

9. Cut plastic straws to 6" lengths. (Length may vary depending on desired distance from top to bottom of carousel.) Place equal distance around base and 3/4" to 1" in from outside edge. Glue to hold. Allow to dry and then glue straws to inside roof of carousel.

10. Glue a horse or animal onto every other pole --- one up and one down for carousel effect. Don't get dizzy, 'cause round and round you go! Carousel horses or animals may be cut from old cards, magazines, etc.

Place on fold

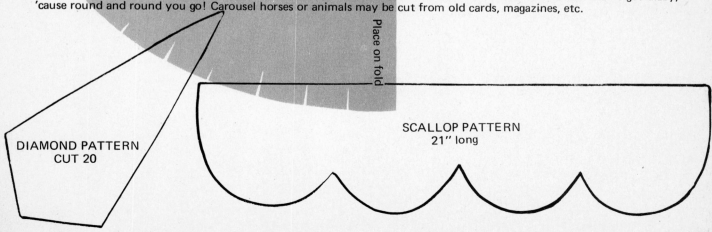

DIAMOND PATTERN
CUT 20

SCALLOP PATTERN
21" long

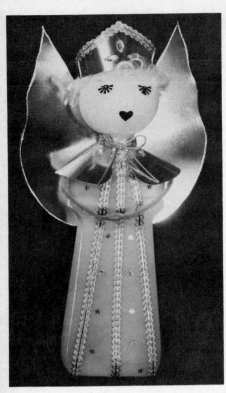

Golden Angel

MATERIALS NEEDED:

1 qt. size detergent bottle - body
1 sheet gold deco foil
1 gold chenille stem - arms
Curly angel hair
White glue

3″ White Foam ball - head
Gold Paper Lace
13″ gold chatelaine cord - bow
Sequins for eyes and mouth
Jewels and sequins for trim

DIRECTIONS:

1. Force White Foam ball over neck of bottle. Use a dab of glue to make it stay better. Pin sequins to face for eyes and mouth. Glue a small amount of curly angel hair to top of head for hair.

2. Cut collar, wings, and crown out of deco foil. Bring collar around neck and fasten with two pins in the front. Place wings to the back and force pins through the wings into the bottle. Overlap ends of crown making a circle. Staple or glue to hold. Glue a strip of Gold Paper Lace around top edge of crown. Place on top of head and fasten with pins.

3. Decorate front of angel with 3 strips of Gold Paper Lace and glue a few tiny stars in between. Glue a strip of the paper lace around the neck of the collar. Make a bow of the chatelaine and glue to front of neck for a tie. Decorate crown and collar with jewels.

4. For arms, cut one chenille stem in half. Put a dab of glue on one end of each piece of stem and force up under collar for arms. Glue to hold. Now bring other ends around to front and loop them together.

CROWN PATTERN

PLACE ON FOLD

PLACE ON FOLD

COLLAR PATTERN

Feather Angel

This fluffy angel is made with a detergent bottle body and 4″ White Foam cone arms. The arms are glued to the sides of the bottle with white glue.

The ready-made head is glued to the top of the bottle. Overlapping rows of floral feathers are glued to the body and arms starting at the bottom. The feathers are placed so that the tips curve up.

The collar is made of a row of feathers around the neck with the feathers curving down.

Perky Santa

MATERIALS NEEDED:

1 detergent bottle - painted red
3″ White Foam ball - head
9″ white celtigal - hat trim
2 moving eyes -1/2″
White glue

4″ x 5″ white felt - beard
6″ square red felt - hat
1 red Xmas ball - nose
Pins

DIRECTIONS:

1. Remove cap from bottle. Force the stem of White Foam ball over the neck. Use a dab of glue to make it stay secure.

2. Force stem of Xmas ball into head for a nose. Glue eyes to face. Cut a mustache and beard out of white felt using patterns given. Pin to face just below the nose. Slip a small piece of red felt in under the beard to fill in the hole between the mustache and beard. This makes the mouth.

3. Cut the cap out of red felt as in Fig. 1. Pin straight edges together making a pointed cap. Pin to top of head. Pin and glue celtigal braid around bottom edge of cap.

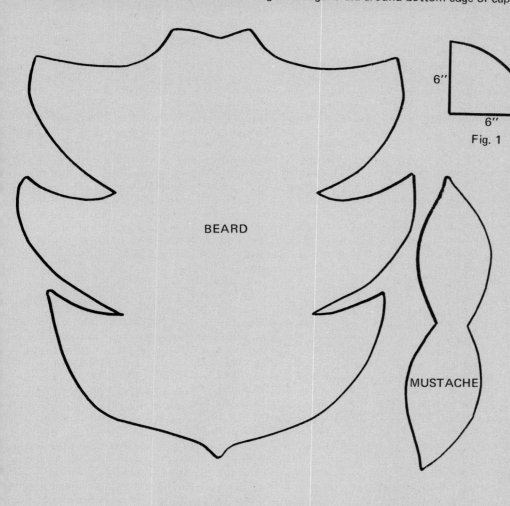

6″

6″

Fig. 1

BEARD

MUSTACHE

Feather Santa

Attach the White Foam head to the detergent bottle as in the Santa above. Glue eyes and nose into place. Add hat made of red and white felt with a small bell secured to the tip.

Cover entire bottle with red feather fluffs. Make sure the bottle is completely covered. The Santa's beard is made of angel hair.

The black belt (made from felt, ribbon, or construction paper) has a gold or silver paper buckle.

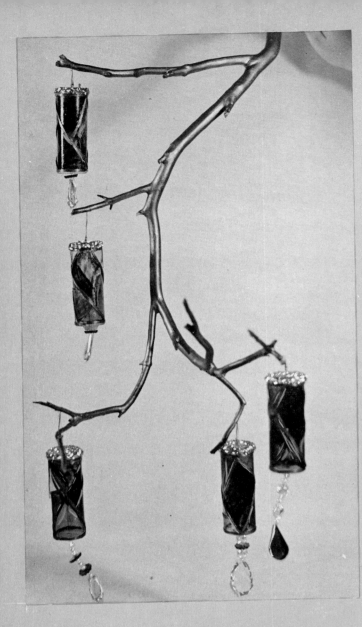

Christmas Ornaments

Here's a clever idea for using the plastic pill bottles you've been saving. These bottles were decorated in bright reds, greens, blues, & yellows, giving a stained glass effect.

MATERIALS NEEDED:

Plastic pill bottles
Lead stripping
Lead adhesive
Transparent paint

Glitter
Paint brush
Glass beads & prisms
Fine wire

DIRECTIONS:

1. Cover one side of the lead stripping with lead adhesive. You may do a long piece at a time and then use it as you need it. Let adhesive dry 10 to 15 minutes or longer before using. This makes it tacky and much easier to handle.

2. Make a design on the bottle with lead strips. Cut lead with razor blade where ever two pieces meet. Let dry thoroughly.

3. Paint in between the lead strips with the transparent paint making it look like stained glass. Let dry. Make a hole through the lid and in the bottom of the plastic bottle with a hot needle. Heat the needle over a flame and plunge through the plastic while it is hot.

4. Thread a prism and beads on a fine wire. Bring the wire up through the hole in the bottom of the bottle and then through the lid. Make a loop at the top for a hanger. Cover lid with glue and then sprinkle with glitter.

NOTE: Pill bottles may be deocrated with many different materials such as sequins, jewels, pearls, braid, sea shells, etc. They may also be painted with pearl lacquer before decorating.

Candle Holder

This is a good youth project or an idea for patio use. A turquoise colored bottle was used with a yellow candle.

MATERIALS NEEDED:

1/2 gallon plastic bleach bottle
Gold braid
Decoupage or pictures from old greeting cards
White glue

DIRECTIONS:

1. Cut top half of plastic bleach bottle off at the seam line just below the handle. Warm bleach bottle, by filling with hot water, to make cutting easier.

2. Glue gold braid around top spout, bottom edge, and on handle.

3. Decorate with decoupage or pictures cut from old greeting cards. Place candle in top spout.

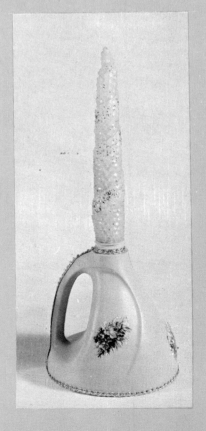

Basic Doll

MATERIALS NEEDED FOR BASIC DOLL:
Small plastic bottle
5/8" wooden bead - head
Tissue paper
White glue
Chenille stem

DIRECTIONS FOR BASIC DOLL:

1. To make the top of the bottle taper more to a point, place a chenille stem down through the neck and wrap tissue paper around the top and neck of the bottle until it has a nice cone shape. Glue to hold. This keeps the shoulders of the doll from being too square.

2. Force the chenille stem up through the hole in the bead head. Paint face on bead or you may buy wooden beads with faces already painted on them.

3. To make the robe, cut a half circle of tissue paper the same length as from the neck to the bottom of bottle. Wrap around doll and secure in back with a bit of glue.

4. For sleeves, cut another half circle of tissue paper making it the length you want the sleeves. Cut in half. Make little cones out of these pieces of tissue paper. Glue where edges overlap. Glue these to both sides of the doll.

Bride

Pinch the paper robe in around the neck of the bottle to make a narrow waist. Twist a piece of chenille stem just below the head for arms. Next bring a narrow piece of net around the neck and over the shoulders like a shawl. Fasten at the waist with thread or fine wire.

Make an overskirt of net by gathering it around the waist. Bend arms around so they meet in the front and twist together. Glue a bit of net and a flower to the top of head for veil. Make a tiny bouquet of two little pieces of ribbon and a few tiny flowers and glue to hands.

Graduate

To make the graduate, make basic doll with black robe and cut a circle of white paper and glue around neck. Glue a square of heavy black paper to the top of head.

Madonna

For the Madonna, make the basic doll body. Dip a strip of tissue paper into water which has a little glue added. Wrap tissue around the doll making it fall into folds. Trim off the bottom if it is too long. Take another strip of wet tissue paper and place it over the head. Bring ends down over the shoulders and arms. When doll is completely dry, it may be painted gold.

Angel

For the angel, cut wings out of heavy gold paper or base metal and glue to the back of the doll. Glue a little circle of gold paper to top of head for crown.

Snow White and the Seven Dwarfs

This idea is a modified version of the old "peek-boxes" of yester-year. The scenes can be as elaborate and minute as you wish. (This is a fun way to use and display miniature collections.)

Take three bottles of different shapes and sizes. Place them together with the handles to the back. Cut out the front as in picture. Now staple bottles together.

Fill bottom of bottle with White Foam and cover with green foil, paper, or moss.

Cut pictures of Snow White and Dwarfs out of magazine or story book and make scene inside bottles. Use artificial flowers and plants to complete scene.

Decorate bottles with Gold Paper Lace, jewels, sequins, beads, and corsage pins. The pointed tops are made out of paper drinking cups. The bottoms are cut off and turned up.

A Summer Day

MATERIALS NEEDED:

Gallon size bleach bottle
Gold Paper Lace
3" sq. White Foam - 1" thick
White Glue

Tiny birds, chickens, & rabbits
Spring flowers
Moss or angel hair

DIRECTIONS:

1. Cut opening on the side opposite the handle. It should be about 5-1/2" high and about 6-1/2" across. Trim edge with Gold Paper Lace.

2. Cut a little house out of the White Foam. Decorate house with Gold Paper Lace. Place house to the back of the bottle and glue to hold.

3. Cover floor with moss or angel hair. Make scene with little sprigs of flowers, birds and rabbits. Fasten two tiny birds to the outside of the bottle. See picture. Glue a few flowers and a tiny bird to neck of bottle.

Magic
With
Tin Cans

Introduction

Tin can crafting is an expressive creative art that is often useful and always extremely interesting and decorative.

No special talent or extraordinary artistic ability is necessary in order to find yourself producing items that exhibit professional quality. It is merely necessary to practice and develop a few techniques that are easy to learn.

The tools used in tin can crafting are simple and inexpensive. Material costs, for the most part, are non-existent except for the paint, bangles and jewels needed for ornamentation. It must be admitted, however, that you might find yourself shopping for containers rather than food when you are in the market.

Methods of measuring, cutting, curling and shaping of tin cans will be described, as well as the assembly of individual forms with others; to make sunbursts, plaques, candle-holders, medallions, Christmas trees, wreaths, and many other interesting and beautiful designs.

We also encourage the tin can hobbyist to use his own imagination and ingenuity in originating cutting designs and curling variations as well as varying the combinations that are assembled in this book, thus forming beautiful creations that depict his own approach to this interesting craft.

May you have many hours of enjoyment with this interesting but different craft which allows the amateur to create with professional results.

Tools, Equipment and their uses

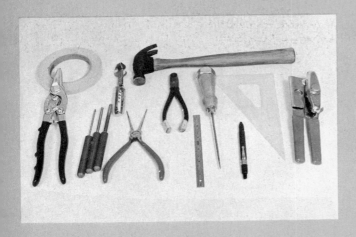

The basic tools and equipment for tin crafting are:

1. Straight cut shears with spring return or compound leverage. (Serrated cutting edges are preferred because the cuts are cleaner and not so sharp and hazardous to touch.) These are used for almost all cuts made in this book.

2. An awl or ice pick. These are used to punch holes in the tin cans and in some instances to mark the tin.

3. Six-inch metal scale or ruler. This is used primarily for measuring up from the rim of the can. It can be used to divide spokes into strips if the template is not used.

4. Special cutting tools. No. 1, (1/8 inch diameter), No. 2, (3/16 inch diameter), and No. 3, (1/4 inch diameter). These are best and easiest for all curling operations of strips, etc.

5. Long nose or needle nose pliers. The pliers are used to bend and shape the tin strips. They can also be used to curl strips if you do not have the special curling tools.

6. Gloves - These are to prevent cuts when cutting and handling the tin cans with the hands.

7. Marker - Used to scribe cuts, etc. on cans. A grease pencil is good.

8. Hammer - Used to flatten the can to the rim when making spokes and to drive the awl or ice pick through the tin.

9. A good quality can opener. This is used to remove the lids and rims from the cans.

10. Old fashioned rip type can opener. This is used to cut the cans around. This is done when it is desired to get two useable halves from a can, each with a rim.

11. A serrated knife for cutting various materials such as White Foam, etc.

12. Wire cutters - Used for cutting chenille stems, wire, etc.

13. A cloth tape measure - Used to measure around cans.

14. A block of wood. This is used to punch holes on and to pound on in general.

15. Scotch tape or masking tape. Used to stick the paper around the cans for marking.

16. Bond cement, clear epoxy, wire and sheet metal screws - These are used for fastening.

17. A small sharp chisel - This is used for slitting can lids in the center when necessary.

18. Last but not least - Cans, cans and more cans of every size and shape.

A few additional tools that might be of use are a good quality kitchen shears, splicer shears or electronic shears, a fine round nose pliers for curling, and a compass or divider for measuring and marking. A proportional divider is of extreme use if available.

The few tools that must be bought should be of reasonably good quality because good tools make the job more fun and much easier.

General Techniques

SPOKES

Step One: Preparation

The can should be washed and clean inside and outside. Most label glues will chip off when dry, but if difficulty is encountered the glue can be removed with nail polish remover or acetone. This will remove ink also.

Remove the lid and top rim from the can. This is easily done with the can opener. See photo 1.

Photo 1

Step Two: Division of the can.

The most common divisions used in the making of spokes are four, eight and sixteen, although other divisions are used. The easiest way to divide a can into any of the above numbers is: Wrap a piece of paper of slightly narrower than the can is high around the can starting just to the side of the seam and cut to length so that it finishes just to the other side of the seam leaving about a 3/16 inch gap with the seam between. See photo 1. Remove the paper and fold in half twice for four divisions, fold in half three times for eight divisions and fold in half four times for sixteen divisions. See photo 3.

Replace the paper around the can in the position described above, tape in place and mark at the top and bottom of each crease with grease pencil. See photo 4. Remove the paper and connect these marks with a ruled line. See photo 5.

Photo 2

If it is desired to divide the can into a different number of spokes the easiest way is to wrap a cloth tape measure around the can and divide the measurement by the number of divisions that are desired. The tape measure should be wrapped around the can in the same manner as the paper was in order to eliminate the seam from the measurement. Refer to photo 2. Again, lines should be ruled on the can to insure straight cuts.

Templates are furnished in this book for the can sizes that are called out in the individual designs. These can be used to divide the cans into spokes and strips. See photos 6 and 7.

Photo 3

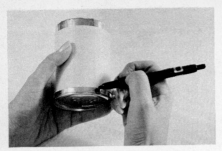

Photo 4

Photo 5

Photo 6

Photo 7

Step Three: Length of spokes and depth of cuts.

If the spokes are to be shorter than the height of the can, measure the prescribed distance above the bottom rim of the can and mark around the can. Also, if the cuts are not to be all the way to the bottom rim of the can, measure the prescribed distance above the bottom rim and mark around the can to define the depth of cut. See photo 5.

Step Four: Cutting

Start at the seam and cut down each side of the seam to the depth that the spokes are to be cut and remove the seam by bending back and forth until it snaps off. Cut down each ruled line to the prescribed depth and, if the spokes are to be shorter than the full height of the can, trim each to the proper length. See photos 8 and 9. Spokes are left in position as shown in photo 9 for crowns. For medallions, sunbursts, centerpieces, etc., the spokes are bent out flat so that they are parallel with the bottom of the can.

STRIPS

Step One: Measuring and marking.

Each spoke is divided into a number of equal parts in order to form strips. This is done by measuring the width of the spoke and dividing this measurement by the number of strips wanted in each spoke. Example: If a spoke were 1 in. wide and eight strips were wanted - 1 in. divided by 8 = 1/8 in. width for each strip. This dimension would then be measured and marked across each spoke to give the eight equal strips.

The other method that can be used for dividing spokes into strips is with the templates given in the book providing the cans used are the size called out in the directions. See photo 7.

Step Two: Cutting

Use the straight cut shears to cut each spoke into strips. Cut each strip either to the bottom rim or to the line marked on the can that defines the depth of cut. See photo 10. Lines can be ruled on the can to aid in cutting the strips straight if necessary.

Step Three: Curling

The strips are curled with the special curling tools or long nose pliers. To curl, take hold of the end of the strip with the tool used and twist or rotate the tool. See photos 11 and 12. The main things to watch in order to do a good job curling are: 1. Take hold of the strip so that the tool makes a right angle with the strip. This enables the strip to be rolled neatly on top of itself. 2. Be careful that the end of the strip does not protrude beyond the tool. This is so that the rolled strips are not egg shaped. 3. Maintain enough tension on the tool so that the strip rolls smoothly and evenly and does not have a tendency to kink.

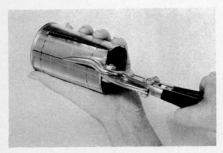

Photo 8

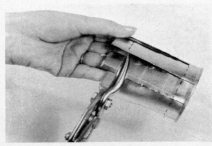

Photo 9

Photo 10

Painting

If the cans are to be painted, normal precautions used when painting any surface should be followed. The surface should be clean and free of grease and oil.

Lacquer or acrylic base paints seem to work best on tin cans. The paint can be either brushed or sprayed. Glass stain also works very well. See photo 13. If intricate surfaces are to be painted the spray paint works best.

If it is desired to soften the color of the painted surface and give a unique finish, a pearlized spray can be put over the paint.

Division and cutting of lids

Three templates are furnished for the division of lids. See below. Each template is used to divide a can lid into a number of equal parts. Template No. 1 is used for 8, 16, or 32; No. 2 for 6, 12 or 24; and No. 3 for 5, 10 or 20 parts. The template is marked with the maximum number of divisions that can be gotten. To get less parts, simply use every other or every third line.

Center the template used on the can lid and mark at the end of the dividing lines used for the number of divisions desired. See photo 14. Rule a line that connects the marks on opposite sides of the lid and passes through its center. See photo 15.

Use a compass to draw a one inch diameter circle around the center of the lid. This is to define the depth of the cut that is made. See photo 16. Cut down each ruled line to the depth marked by the circle.

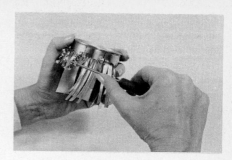

Photo 11

Photo 12

Photo 13

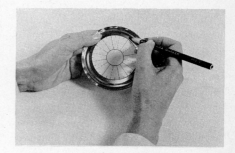

Photo 14

Template No. 1
8 - 16 - 32 equal parts

Template No. 2
6 - 12 - 24 equal parts

Template No. 3
5 - 10 - 20 equal parts

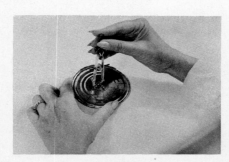

Photo 16

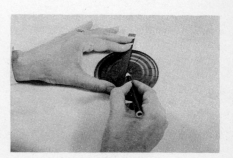

Photo 15

137

Jeweled Christmas Tree

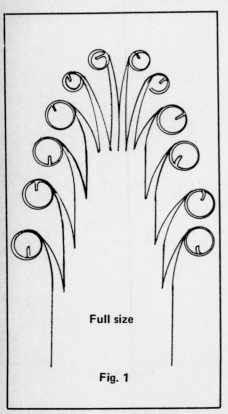

30″

Materials:

1 - 2½″ x 10″ White Foam Disc
1 - 3″ x 2′ White Foam rod
1 - 2″ x 1½′ White Foam rod
1 - ½″ x 3′ dowel stick
1 - 1 gallon paint thinner can - rectangular
11 - one gallon cans
2 - 5 inch diameter, 2 lb. coffee cans
3 - 4 inch diameter, 42 oz. juice cans
9½ doz. 30 mm Christmas balls
1½ doz. 20 mm Christmas balls
3 - 7½ inch bead sprays
100 crystal drops - clear
100 - 2 inch head pins - hangers
22 - Gold chenille stems
1 doz. chenille stems - fasteners
Gold spray paint

Step One: Division and sizing of cans

Refer to the instructions in the General Technique section, page 135, to:

1. Divide the nine one gallon cans into sixteen equal spokes, the two 5 inch diameter, 2 lb. coffee cans into 14 equal spokes and the three 4 inch diameter, 42 oz. juice cans into 12 equal spokes. Rule lines on each can to insure straight cuts.

2. Cut each can to the bottom rim following the ruled lines.
3. Bend the spokes of each can flat at the bottom rim.

Locate the center of each can. This can most easily be done by laying a piece of paper across the can as near to center as possible and cutting the paper so that it fits from one side of the rim to the other. Fold the paper in half, place it against the rim and make a mark at the end. Move part way around the can and repeat. Where the two lines cross marks the center of the can. Punch a hole in the center of each can.

Leave two of the one gallon cans full size. Measure from the center of these cans to the end of the spokes in order to establish a starting dimension. Measure from the center of the third can and mark it 1/4 inch shorter than the first two cans. Repeat with all cans marking each one progressively shorter than the last by 1/4 inch. Number each can as it is marked so that when assembled, uniform taper of the tree can be maintained. Trim the spokes of each can to the length marked.

Step Two: Cutting and curling of strips

Three templates are furnished, one for each size of can that is used. These are used to divide the spokes into equal strips and to mark the depth of cut of each strip.

Use the proper template to divide the spokes of each can into strips and mark the depth of cut of each strip. Cut the strips to the width and depth marked.

Full size

Fig. 1

Fig. 1 is a full size illustration of a spoke with its strips curled. Curl the strips of each spoke on each can as shown. The No. 2 curling tool, (3/16 inch diameter), is used to curl the four center strips of each spoke and the No. 3 curling tool, (1/4 inch diameter), is used to curl the rest of the spokes. Long nose pliers can also be used to curl.

Step Three: Preparation for assembly

Use the template that was used to divide the spokes into strips to locate and mark two holes on every other spoke. Use an ice pick to punch small holes where marked. These will be used to fasten the Christmas balls.

Draw a short line through the center of each can. Measure 3/4 inch each side of center along this line and punch a hole. These holes will be used to secure cans so that they will not turn after the tree is assembled.

Punch a hole in each can large enough to insert the 1/2 inch dowel stick through. This can most easily be done by using a small sharp chisel or screw driver to make about six cuts approximately 3/8 inch long to the center hole in the can. See Fig. 2.

Refer to the procedure in the General Technique section, page 137, to prepare the cans and paint them gold.

Step Four: Additional spokes and top-piece of the tree

Divide and cut two one gallon cans into sixteen equal spokes as described in Step One. Cut each spoke off at the bottom rim of the can. Twenty-seven of these spokes are used. Cut ten spokes to a length of 6-1/4 inches, 8 to a length of 6 inches, 8 to a length of 5-3/4 inches and 1 to a length of 4-1/4 inch.

Follow the instructions in Step Two to divide each spoke into strips, cut the strips and curl them. Trim the other end of each spoke as shown in Fig. 3. Punch two holes as shown in Fig. 3 to fasten spokes when assembling tree. Also, locate and punch holes for mounting Christmas balls as was done in Step Three. This should be done to half of each length spoke that was cut.

Remove the top and bottom lids from the rectangular shaped one gallon paint thinner can. Use the tin shears to cut the can open at the seam. Lay the can out flat and remove the top and bottom rims. Measure and mark five spokes the same width as the spokes from the round one gallon cans and 10-1/2 inches long. Cut these spokes out. Use pattern 1 to mark the shape of the spokes. divide each spoke into strips, mark the depth of cut of the strips and locate the holes that are used to mount the spokes on the tree. Cut and curl these strips exactly the same as described in Step Two. Punch the holes for mounting.

Paint all spokes and the top-pieces gold.

Step Five: Assembly of the tree

Paint the 10 inch diameter White Foam Base, the 3 inch diameter pole and the 2 inch diameter pole gold and allow to dry. Sharpen the 1/2 inch diameter dowel stick to a pencil like point. Locate the center of the 10 inch base and stick the dowel stick through the base making sure that it is straight up, i.e., perpendicular to the base. The sharp end of the dowel stick should be up.

Cut nine 1-1/2 inch long spacers out of the 3 inch diameter White Foam pole. Cut four 1-1/2 inch long and four 1-1/4 inch long spacers out of the 2 inch diameter White Foam pole. Cut these spacers as near to square as possible.

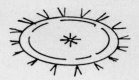

Fig. 2

Fig. 3

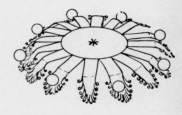

Fig. 4

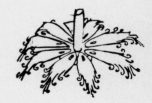

Fig. 5

Fig. 6

Fig. 7

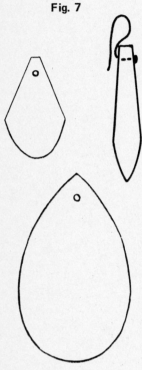

Fig. 8

Use crystal drops of your choice. See suggestions shown above.

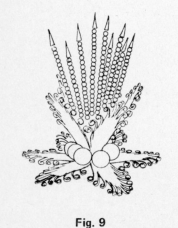

Fig. 9

When the tree is assembled the inside of the cans are up, i.e., the top side of the assembly.

Remove the hanger or stem from a 30 mm Christmas ball. Bend a 2-1/4 inch piece of gold chenille stem and insert it from the bottom through a spoke with the holes between the curled strips. Twist the stem above the spoke, leave the ends spread in a "V" shape and place the Christmas ball on the stem. Do this around the can with each spoke that has holes punched in it for ornaments. Bend each spoke down at the bottom of the deepest cut strip so that the end of the strips are 3/4 to 1 inch below the body of the spoke. See Fig. 4. Do all cans in this manner.

Force the point of the dowel stick through the center of a 3 inch diameter spacer and slip the spacer down the dowel stick until it rests on the base. Put the can marked "1" on the dowel stick and slide it down until it rests on top of the spacer. Insert a piece of chenille stem approximately 1-3/4 inches long into the holes to either side of the center. The pieces of chenille stem should be inserted into the spacer below to a depth equal to about half their length. Put another 3 inch diameter spacer on the dowel stick and push it down until it is flush on top of the can and the chenille stems are driven into the spacer. Next place the can marked "2" on the dowel stick and slip it down until it rests on the spacer. Rotate the can so that its spokes are between the spokes of the number "1" can. Insert the chenille stems as above, then another spacer and the can marked "3" and so on repeating the above procedure for each additional can. The 3 inch diameter spacers are used for the first nine cans and the 2 inch diameter spacers are used for the rest of the tree. The tree at this point will be done through can 14. When can size and number of spokes change in the assembly, it will not be possible to make the spokes of the upper can match with the voids in the lower can, but a fairly good match can be attained.

Push the dowel stick through the center of one of the 1-1/4 inch high 2 inch diameter spacers and slide the spacer down until it is flush on top of can 14. Bend a 1-1/2 inch piece of chenille stem into a "U" shape and insert in the mounting holes on each spoke. Place 30 mm Christmas balls on the five 6-1/4 inch spokes that were made and 20 mm Christmas balls on the four 6 inch and four 5-3/4 inch spokes that were punched. Bend the ends of the spokes down as was done with the spokes on the can. Place the ten 6-1/4 inch spokes around the 1-1/4 inch spacer on top of can 14. Fasten the spokes by pushing the "U" shaped piece of chenille stem into the White Foam spacer. Follow the same basic pattern with these spokes as is followed throughout the tree with the cans, i.e., every other spoke with a Christmas ball and the end of the spokes lining up with the empty space between spokes of the section immediately below. Place another 1-1/4 inch spacer on the dowel stick and repeat with the eight 6 inch spokes; then another spacer and the eight 5-3/4 inch spokes. See Fig. 5.

Place the last 1-1/4 inch high, 2 inch diameter spacer on the dowel stick and cut the dowel stick off flush with the top of the spacer. Shape the top-pieces of the tree as shown in Fig. 6 and insert a piece of chenille stem bent in a "U" shape into the mounting holes of each top-piece.

Mount the top-pieces as shown in Fig. 7 and fill in the empty space left between the horizontal strips with the 4-1/2 inch spoke.

Step Six: Completion of tree

The bead sprays come in groups of five stems. Separate each spray so that you have fifteen stems. Insert twelve of these stems into the top White Foam spacer around the dowel stick and in the middle of the top-piece.

Insert the 2 inch head pins through the holes in the crystal drops and form into hangers. See Fig. 8. Distribute the crystal drops around the tree and hang them from the curled strips. String six 20 mm Christmas balls on a wire and space as shown in Fig. 9. Tighten the wire to hold in place.

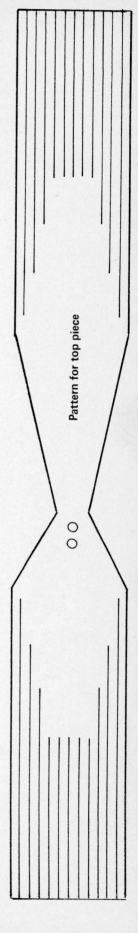

4 inch diameter can
Template

5 inch diameter can
Template

1 gallon can
Template

Pattern for top piece

Fancy Fantasy Bubble

Materials:

1 - 5 inch diameter 2 lb. coffee can - plastic lid
1 - 5 inch diameter clear acetate fantasy bubble
1 - small brush Christmas tree - green - 4 inch
1 - Christmas bead string - 1/2 inch diameter beads
Assorted colors of seed beads
Scraps of foil wrapping paper
White Foam scraps
Glitter glue
White glue

Step One: Tin Collars

The unit is comprised of two identical Tin Collars.

These are made from the 5 inch diameter 2 pound coffee can. The can is to be cut in half, (around), using an old fashioned rip type can opener. Start the cut by punching a hole through the can near the seam with the ice pick. Insert the can opener in the punched hole and proceed to cut. See Fig. 1. Remove the bottom lid from the can. (This should not be done until the can is cut around.)

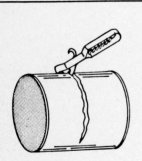

Fig. 1

Fig. 2

If the can has a painted label, it should be painted gold before cutting. This is to be sure that the color of the can is well covered when the unit is finished.

Refer to the instructions in the General Technique section, page 135, to:

1. Measure 2-1/2 inches above the rim of each half of the can and mark around the can. This denotes the length of the spokes.

2. Divide each half of the can into sixteen equal spokes, cut each spoke to the bottom rim and trim each spoke at the 2-1/2 inch point located above.

3. Divide the spokes of each can into twelve equal strips and cut each strip to the bottom rim.

Straighten out all the strips of each half of the can so that you have two flat pieces. Place the two halves of the can together with the smooth side of the rims facing one another. Move until the seam on each half of the can matches identically, i.e., seam against seam and lines up edge of seam to edge of seam. Use fine wire to fasten the two halves of the can together in this position. This is done so that the two sides of the frame will line up strip for strip when formed.

Start at the seam, count three strips and leave them as they are. Bend the next six strips down so that they point in the normal direction of a can side-wall, leave the next six up, bend the next six down and so on around the can. Finish with the last three next to the seam pointing the same as the first three. Turn the unit over, start at the same point next to the seam and bend the strips in the same manner. Make sure that the strips on each side match identically. If necessary, move halves slightly so that they do line up strip for strip. See Fig. 2.

Lay a piece of half inch dowel stick over the six strips from each half of the can that are bent in the normal direction of the can sidewall. Form three strips from each adjacent group on each side of the can around the piece of dowel stick. Do this with all groups around the can. See Fig. 3

Curl the center two strips that were formed around the dowel stick to a length of approximately 1-1/2 inch. The two next to center should be curled to a length of 1-1/4 inch and the outside two of each group to 1 inch. See Fig. 4.

Start at the same end of the groups on each side of the unit that were bent down, count three strips and bend the third strip on each side up. Do this with each group that was bent down. These strips will be used to fasten the two halves of the unit together. There are now five strips left in each group that are pointing in the normal direction of the can side. Start with the Center strip of each group and curl to a length of one inch. Curl the two next to center to approximately 3/4 inch and the outside two of each group to about 1/2 inch. Do each group that was bent down on each side of the unit in this manner. See Fig. 4.

Step Two: Bubble scene

Spray the four inch Christmas tree with glitter glue and sprinkle with the assorted color seed beads. Use the scrap White Foam and foil paper to make miniature Christmas packages to place around the tree.

Place white glue on 1/2 the tree base and glue on one part of the fantasy bubble so that the base is half on and half off the edge. See Fig. 5. After the glue holding the tree has had time to dry, place glue on part of the packages and glue them around the tree base in the bubble. Avoid getting too much cement on the acetate bubble. Put glue on the remaining packages and glue them to the tree base and to one another so that they are freely suspended and the other half of the bubble will fit.

After the cement is dry, place Bond cement on the edge of the fantasy bubble and fasten with clothes pins until fused. After the bubble has been allowed to fuse, trim around so about 1/8 inch lip is left. See Fig. 6.

Step Three: Painting and assembly

Remove all the wires fastening the two collars together except one at the seam that will act as a hinge. This one is left so that the alignment of the strips will not be disturbed.

Repaint the collars with gold paint. Make sure that collars are completely covered and that there are no runs. Allow to dry completely before assembly.

The group on each side of the seam on each side of the unit are used as legs. Place the bubble scene between the two collars with the tree base directly above the seam. Close the collars. Start at the top of the unit and thread the strips that were bent part way up into the area where the strips were formed around the dowel stick. Slip a 1/2 inch diameter Christmas bead over the two strips and curl each strip almost to the top of the bead. Do this all the way around the unit. See Fig. 7.

Flare the bottom two groups that are to be used as feet and even them up so that the unit sets level. See Photo.

Fig. 3

Fig. 4

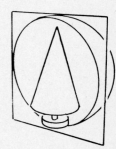

Fig. 5

Fig. 6

Fig. 7

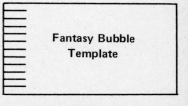

Fantasy Bubble
Template

Garlanded Lamp

Materials:

2 - 2-5/8 inch diameter, 11 oz. soup can
1 - 3-1/8 inch diameter, 1 lb. can
1 - 4 inch diameter, 1 lb. shortening can
1 - 3 inch diameter can lid
12 doz. 5 mm pearls with holes
1 - 8-1/2 inch high x 3 inch bottom dia. clear glass
    chimney. Bead Wire
1 - 4-1/2 inch candle
1 - Sheet metal screw
Gold paint

Step One: Top Collar

Use a 2-5/8 inch diameter can to make the Top Collar. Cut the bottom lid out of the can. (This is the difference between a collar and a crown.)

Refer to the directions in the General Technique section, page 135, to:

1. Measure 1-1/2 inch above the bottom rim of the can and mark around the can to denote the length of the spokes.
2. Divide the can into sixteen equal spokes, cut each spoke to the bottom rim and trim each spoke at the 1-1/2 inch point marked above.
3. Divide each spoke into five equal strips and cut each strip to the bottom rim of the can.

Use the long nose pliers to turn each strip of each group at right angle to the can, and to squeeze or gather the five strips of each group 3/8 inch above the bottom rim of the can. See Fig. 1.

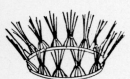

Fig. 1

Fig. 2

Fig. 3

Use the No. 1 curling tool, (1/8 inch diameter), or long nose pliers for all curls on the Top Collar. Curl the center strip of each group to a length of 1 inch. The strips to either side of center should be curled so that they are just slightly shorter than the bottom of the curl on the center strip and the outer two strips should be curled just slightly shorter than the bottom of the curl of the strips next to center. Curl each group in this manner and flare each group outward so that it fits the contour of the chimney. See Fig. 2.

Step Two: Top Crown

Use the 3-1/8 inch diameter can for the Top Crown. Refer to the directions in the General Technique section of the book to:

1. Measure 2 inches above the bottom rim of the can and mark around the can to denote the length of the spokes.
2. Divide the can into fourteen equal spokes, cut each spoke to the bottom rim of the can and trim each spoke at the 2 inch point marked above.
3. Divide each spoke into five strips and cut each strip to the bottom rim of the can.

Follow the instructions given in Step One—Top Collar—to:
1. Turn and gather the strips of each group.
2. Curl the center strip of each group to 1-1/4 inch length and the other strips of the group progressively shorter as described.

144

3. Flare the groups outward so that the inner edge of the curl of the center strip is approximately 1/2 inch outside the inner edge of the bottom rim of the can. See Fig. 3.

Step Three: Centerpiece

Use the four inch diameter can to make the centerpiece. Refer to the instructions in the General Technique section, page 135, to:

1. Measure 2 inches above the bottom rim and mark around the can to define the length of spokes and strips.
2. Divide the can into twenty-four equal spokes, cut each spoke to the bottom rim of the can and trim each spoke at the 2 inch point marked above.
3. Divide each spoke into five equal strips and cut each strip to the bottom rim of the can.

Bend all the strips down so that they become an extension of the bottom lid of the can. Turn and gather the strips of each group as described in Step One - Top Collar. Curl and form each group exactly the same as for the Top Crown of Step Two. See Fig. 4.

Step Four: Foot Crown

Use a 2-5/8 inch diameter can to make the Foot Crown. Refer to the instructions in the General Technique section of the book to:

1. Measure 1 inch above the bottom rim of the can and mark around the can to define the depth of cut. Measure 2 inches above the bottom rim of the can and mark around the can to denote the length of spokes.
2. Divide the can into 12 equal spokes, cut each spoke to the 1 inch point marked above and trim each spoke at the 2 inch point marked.
3. Divide each spoke into five equal strips and cut each strip to the 1 inch point located above.

Turn, gather and curl the strips of the Foot Crown exactly the same as those of the Top Collar with the exception that the 1 inch measurement for the length of the center strip is to be made from the bottom of the cut rather than from the bottom rim of the can. Do not flare the groups of the Foot Crown outside the bottom rim of the can. See Fig. 5.

Step Five: Candleholder

Use the 3 inch diameter can lid to make the Candleholder. Locate the center of the can lid and use the template to divide the lid into eight equal parts. Use a compass and draw a circle around the center-point of the lid that is the diameter of the candle that is to be used. This is to define the depth of cut. Cut the can as shown in Fig. 6. Bend the four petal shaped sections up so as to hold the candle. Roll the top of the petals back and curl the six strips between each pair of petals into a decorative cluster. See Fig. 7.

Step Six: Preparation and painting

Punch a hole just large enough to get the sheet metal screw started in the Candleholder, Top Crown, Centerpiece and Foot Crown.

Follow the instructions for painting in the General Technique section, page 137, and paint all parts gold.

Step Seven: Assembly and decoration

Start the sheet metal screw from the top, through the candleholder, Top Crown, Centerpiece and Foot Crown, and tighten until parts are secure.

String the 5 mm pearls on Bead wire. Wrap one string around the Top Collar just above the rim, one around the Top Crown laying on the Centerpiece and one string around the Foot Crown just above the bottom of the strips.

Fig. 4

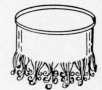

Fig. 5

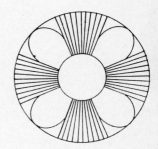

Fig. 6

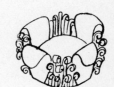

Fig. 7

Collar Template

Top Crown Template

Centerpiece Template

Foot Crown Template

Spanish Candle Holder

Materials:
1 - 4 inch diameter, 1 lb. shortening can
1 - 2-1/8 inch diameter, 6 oz. sauce can
1 - 2-5/8 inch diameter, 11 oz. soup can
1 - Candle holder - screw type
1 - Green votive candle glass - candle holder mount
16 - Emerald beads - 12 mm
16 - Flat shaped crystal beads - 8 mm
Bead wire
Gold paint (spray can)

Step One: Top Crown

Use the 2-1/8 inch diameter, 6 oz. sauce can to make the Top Crown. Refer to the instructions in the General Technique section, page 135, to:
1. Measure 3 inches above the bottom rim and mark around the can to denote the length of spokes. Measure 1/2 inch above the bottom rim and mark around the can to define the depth of cut.
2. Divide the can into seven equal spokes. Cut each spoke to the 1/2 inch point marked above and trim each spoke at the 3 inch point.

Fig. 1

Fig. 2

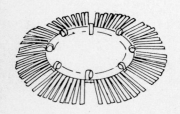

Fig. 3

3. Divide each spoke into eight equal strips and cut each strip to the point 1/2 inch above the bottom rim that was marked.

Start at the seam and use the No. 1 curling tool, (1/8 inch diameter), or long nose pliers to curl the first strip to the bottom of the cut. See Figure 1. Repeat with every eighth strip. This divides the can into seven groups of seven strips each. Curl the center strip of each group to a height of 1-5/8 inch above the bottom rim of the can. Curl the strips on each side of the center strip progressively shorter forming a triangular shape with each group. The groups of strips should flare slightly outside the bottom rim of the can. See Figure 1.

Step Two: Foot Crown

Use the 2-5/8 inch diameter, 11 oz. soup can to make the Foot Crown. Refer to the directions in the General Technique section, page 135, to:
1. Measure 3 inches above the bottom rim of the can to denote the length of spokes and measure 1/2 inch above the bottom rim to define the depth of cut.
2. Divide the can into eight equal spokes. Cut each spoke to the 1/2 inch point marked and trim at the 3 inch mark.
3. Divide each spoke into eight equal strips and cut each strip to the 1/2 inch point marked.

Use the No. 1 curling tool, (1/8 inch diameter), or long nose pliers, start at the seam of the can and curl the first and eighth strips of each group to the bottom of the cut. Curl the second, third, and fourth progressively longer, making the fourth 1-1/2 inches long. The fifth strip is to be curled to the same length as the fourth, and the sixth and seventh progressively shorter to the same lengths as the third and second respectively. See Figure 2. The center two strips of each group, (the fourth and fifth), should be turned at right angles with the rim of the can so as to form more sturdy feet. Repeat for each group of eight.

Step Three: Centerpiece

Use the 4 inch diameter, 1 lb. shortening can to make the Centerpiece. Refer to the instructions in the General Technique section, page 135, to:

1. Measure 1-1/2 inches above the bottom rim of the can and mark around the can. Divide the can into eight equal spokes and cut each spoke to the bottom rim.

2. Divide each spoke into ten equal strips. Cut each strip to the bottom rim. Bend all strips down at the bottom rim so that you have a flat piece.

The can is divided into eighty equal strips. Start at the seam and bend the first strip up, then curl down as shown in Figure 3 using the No. 3 curling tool, (1/4 inch diameter), or long nose pliers. Bend every tenth strip up and curl in this manner. This will divide the can into eight groups of nine strips each.

Use the No. 1 curling tool, (1/8 inch diameter), or long nose pliers and start with the middle strip of each group. Curl this strip to a length of 3/4 inch as shown in Figure 4. Then with the same tool, curl the strip next to the center strip in the same manner as the center strip until it is slightly shorter than the center strip. Without removing the tool, turn until it is at right angle with the rim of the can. Work each way from the center strip curling each strip progressively shorter and turning at right angle. The last strip on each side of the group is rolled to a length of about 1/8 inch from the bottom rim of the can. Do each group of nine in this manner. See Figure 5.

Step Four: Painting

All holes and rough work should be done before the cans are painted. Locate the center of the Top Crown, Foot Crown and the Centerpiece. Use the ice pick to punch a hole that is just large enough for the screw on the candle holder to start into, but smaller than the body size of the screw.

Follow the instructions in the General Technique section, page 137, and paint all pieces with gold spray paint.

Step Five: Assembly and decoration

After the painted parts have been allowed to thoroughly dry, place the Top Crown above and the Foot Crown below the Centerpiece. Line up the holes in the centers of the cans and screw the candle holder in from the Top Crown side. Tighten until all parts are firmly together. See Photo.

To add trim, select one of the strips on the Centerpiece that curls up towards the candle holder; start a wire through the curled strip; string a flat crystal bead, two emerald beads and a flat crystal bead onto it. Continue by threading the wire through the next upright curled strip and repeat the bead pattern. Do this around the Centerpiece. See Figure 6. Other beads and methods of decoration can be used if desired.

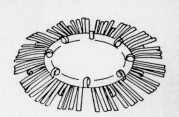

Fig. 4

Fig. 5

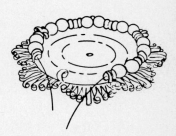

Fig. 6

Top Crown
Template

Foot Crown
Template

Centerpiece
Template

Pink Jeweled Candle Holder

Materials:

1 - 4 inch diameter, 1 lb. shortening can

2 - 2-5/8 inch diameter, 11 oz. soup cans

34 - Pink Dazzle Beads - 8 mm size

16 - Crystal beads - flat shaped - 8 mm dia.

20 - Pink jewels - 4 mm size

9 - Large pink plastic drops - 1-1/4 in. L. x 3/4 inch W.

1 - Crystal votive candle glass

Bead wire

Glue or clear epoxy

Gold spray paint

Sheet metal screw

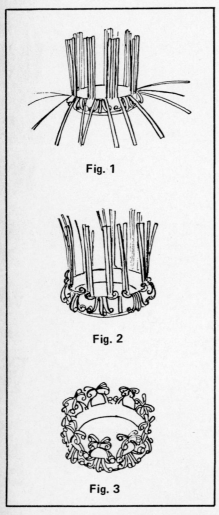

Fig. 1

Fig. 2

Fig. 3

Step one: Top Crown

Refer to the instructions in the General Technique section, page 135, to do the following:

1. Measure three inches above the bottom rim of the can and mark around the can to denote the length of the spokes and measure one half inch above the bottom rim to define the depth of cut.

2. Divide the can into nine equal spokes. Cut each spoke down to the one half inch mark located and trim each spoke at the three inch mark.

3. Divide each spoke into ten equal strips. Cut each strip down to the one half inch mark located above. Start at the seam, count nine strips and cut off the tenth strip at the one half inch point above the bottom rim. Do this to every tenth strip around the can.

Bend the center five strips of each group of nine down at the bottom of the cut. The outside two on each side of the group should be left pointing up. Use the No. 1 curling tool, (1/8 inch diameter), or long nose pliers for all curls on the Top Crown. Curl the center strip of each group up, rolling inward toward the can, until it is even with the bottom rim of the can. Curl the strips to either side of the center strip slightly shorter than the center strip, turn the curled section at right angle to the can with the side away from the strip pointing up and place as shown in Figure 1. Curl the outside two strips that are pointing down to approximately the same length as the previous two strips curled. Place them so that they are slightly above the half inch point above the rim and are just to the point where the strip between groups was removed. See Figure 2. Turn the outside two strips of each group at right angle to the can and form these strips around a piece of 1/2 inch dowel stick. Gather the four strips just above the dowel stick with the long nose pliers, remove the dowel stick and tie at the gathered point with a piece of scrap tin. Curl all four strips to a height of about 1/2 inch above the tie. Curl so that two strips flare forward and two flare back towards the inside of the can. When looking down on the can a "V" shape should be formed by the two strips on each side of the group. See Figure 3.

Step Two: Centerpiece

Refer to the directions for the Centerpiece of the Spanish Lamp, page 147. The Centerpiece of the Pink Candle Holder is identical.

Step Three: Foot Crown

Use a 2-5/8 inch diameter can to make the Foot Crown. Refer to the directions in the General Technique section, page 135, to:

1. Measure 2-1/2 inches above the bottom rim of the can to denote the length of spokes, and measure 1/2 inch above the bottom rim to define the depth of cut. Mark around the can at each measurement.
2. Divide the can into ten equal spokes and cut each down to the 1/2 inch point marked above.
3. Divide each spoke into seven equal strips and cut each strip down to the 1/2 inch point marked above.

Bend the center three strips of each group down. Use the No. 1 curling tool, (1/8 inch diameter), or long nose pliers for all curls on the Foot Crown. Curl the center strip up, rolling towards the can until it reaches to the bottom rim. Curl the strip to either side of the center in the same manner, then turn it at right angle to the rim of the can and place as shown in Figure 4. Repeat for each group.

The outside two strips on each side of the group should be turned at right angle to the can and formed around a 1/2 inch diameter dowel stick. The strips should be gathered above the center of the dowel stick with the long nose pliers. Tie at the gathered point with a piece of scrap tin about 1/8 inch wide. Curl the outside strip on each side of the group to just above the tie and the center two slightly longer. See Figure 5. Do each group in this manner.

Step Four: Preparation and painting

Locate the center of each can and punch a hole in each with the ice pick. This hole should be just large enough to get the sheet metal screw started.

Follow the instructions in the General Technique section, page 137, and paint the cans gold.

Step Five: Assembly and decoration

Place the Top Crown above and the Foot Crown below the Centerpiece. Line up the holes, start a sheet metal screw from the Top Crown side and draw up until parts are tight.

Place the large plastic drops between the "V's" formed by adjacent groups, push them down until they are as near to the bottom of the cut as they will go, and glue in place with Bond cement or clear epoxy. See Figure 6. Place a 4 mm pink jewel in each of the outside curls of the "bent down strips" and glue in place. There are eight strips curled up toward the center of the Centerpiece. Thread bead wire through these curls with an 8 mm flat white bead, three 8 mm pink Dazzle Beads and an 8 mm flat white bead in that order between each pair of upward curled strips around the can. Glue a pink Dazzle Bead in the area above each foot of the Foot Crown. See Figure 7.

Place the crystal votive candle glass in the Top Crown and adjust the feet so that the unit sits level.

Fig. 4

Fig. 5

Fig. 6

Fig. 7

Top Crown Template

Foot Crown Template

Centerpiece Template

Petite Candle Holder

Materials:

1 - 2-1/8 inch diameter, 6 oz. sauce can

1 - 2-5/8 inch diameter, 11 oz. soup can

1 - Blue votive candle glass

28 - Blue Plexi - drops

Bead wire

1 - Sheet metal screw

Step One: Top Crown

The Top Crown is made out of the 2-1/8 inch diameter can. Refer to the directions in the General Technique section, page 135, to:

1. Measure 2-1/2 inches above the bottom rim of the can and mark around the can. This denotes the length of the spokes. Measure 1/2 inch above the bottom rim and mark around the can to define the depth of cut of spokes and strips.

2. Divide the can into eight equal spokes. Cut each spoke to the 1/2 inch point marked above and trim each spoke at the 2-1/2 inch point marked.

3. Divide each spoke into seven equal strips and cut each strip down to the 1/2 inch point marked.

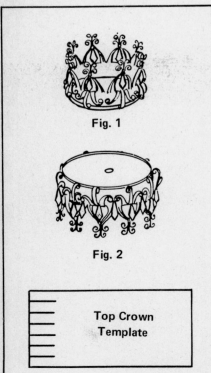

Fig. 1

Fig. 2

Top Crown Template

Refer to the curling instructions in the Foot Crown section of the Pink Jeweled Candle holder on page 149. Each group of the Top Crown is curled the same as the groups in the Foot Crown of the Pink Candle holder. See Figure 1.

Step Two: Foot Crown

Follow the instructions in the Foot Crown section of the Pink Jeweled Candle holder on page 149. The Foot Crown of the Petite Candle holder is identical.

Step Three: Assembly and decoration

The Top Crown and the Foot Crown can either be painted or left natural. The one in the picture is left the natural silver of the cans.

Locate the center of the Top and Foot Crowns. Punch a hole in the center of each with the ice pick. The hole should be just large enough to get the sheet metal screw started. Place the Top Crown on the Foot Crown with the center holes lined up and fasten together with a sheet metal screw.

Use bead wire to hang a Plexi-drop bead from the tie point of each cluster of the Top Crown. Also, use wire to hang a Plexi-drop bead from the bottom of the center strip which is bent up toward the rim of the can rim, and, from the union of the outer strips of adjacent groups of the Foot Crown. See Figure 2.

Place blue votive glass candleholder in Top Crown and adjust curled feet of the Foot Crown so that the piece sets level and flat.

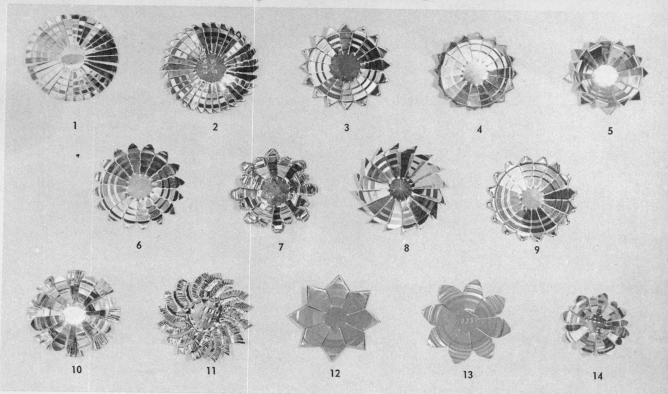

Step One: General

Medallions, in general, are not used by themselves for decorative purposes. Instead, they are stacked into assemblies, decorated and used as building blocks in other projects. The medallions shown and described in this section are used in the wreath and Christmas tree in following sections.

The medallions will be broken into sections by the number of spokes or divisions that they are cut into. Refer to the instructions in the General Technique section, page 137, to divide and cut the lids into any number of parts called out in the individual medallion design. The straight cut shears are used for all cuts, the long nose pliers for all bends, and the No. 1 curling tool, (1/8 inch diameter), or long nose pliers for all curling operations.

Step Two: 32 spoke medallions

Numbers 1 and 2 are 32 spoke medallions. Make them as follows: 1) Divide a lid into 32 parts and cut to center circle. See number 1. 2) This medallion is identical to number 1 except the same corner of each spoke is bent up. See number 2.

Step Three: 16 spoke medallions

Numbers 3 to 11 inclusive are 16 spoke medallions. Make them as follows: 3) Divide a lid into 16 spokes and cut to center circle. Bend the corners of each spoke under as shown in number 3. 4) Medallion number 4 is identical to number 3 except that the end of each spoke is bent up. See number 4. 5) Number 5 is the same as number 3 except that every other spoke is bent up and the ends of all spokes are bent down as shown in number 5. 6) medallion 6 is the same as number 3 except the ends are trimmed as shown in number 6 instead of turned under. 7) Medallion 7 is a 16 spoke medallion which has its ends trimmed, edges of the ends of the spokes feathered with fine cuts and every other spoke bent as shown in number 7. 8) Medallion 8 is a 16 spoke medallion with the end of each spoke cut on a bias as shown in number 8. 9) Medallion 9 is a 16 spoke unit with the end of each spoke curled up as shown. See number 9. 10) Number 10 is a 16 spoke medallion with every other spoke bent up and trimmed as shown. The spoke left down is cut into a series of thin strips part way down. See 10. 11) Looking down on the piece, start at one of the lines on the outer edge of the can and cut in an arc to the point where the next line to the right intersects the circle that defines depth of cut. Do this around the can and put the fine cuts in each spoke as shown in number 11.

Step Four: 8 spoke medallions

Numbers 12 to 21 inclusive are 8 spoke medallions. Make these medallions as follows: 12) Divide a lid into 8 spokes and curl the ends under as shown in number 12. 13) Number 13 is the same as 12 except the ends of the spokes are trimmed as shown in 13. 14) Divide a lid into 8 spokes, cut a strip about 1/3 the width of the total spoke and trim and shape as shown in 14. 15) Divide a lid into 8 spokes, cut 1-1/8" strip off each spoke and curl as shown in number 15. 16) Number 16 is the same as 15 except that 2 strips are cut off each spoke and curled. See 16. 17) Number 17 is an 8 spoke medallion with 3

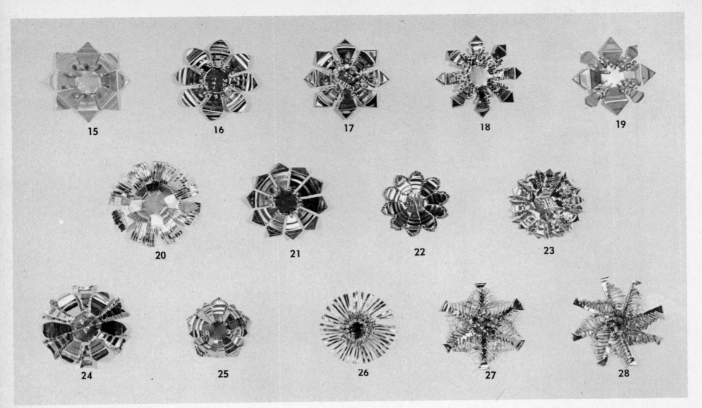

strips cut off each spoke. The ends of the spokes are turned under and the strips curled as shown in number 17. 18) Number 18 is the same as 17 except that there are 3 strips cut off each side of each spoke and curled as shown in 18. 19) Number 19 is the same as 18 except that there are 3 strips cut off each side of every other spoke instead of every spoke. The strips are curled in the same manner. See 19. 20) Divide a lid into 8 spokes, cut strips toward center of can as shown until "V" comes out of middle of each spoke and shape as shown. See 20.

Step Five: 10 spoke medallions

Numbers 21 to 25 inclusive are 10 spoke medallions. They are made as follows: 21) Divide a lid into 10 spokes, cut 1/8" strips off each spoke and curl and shape as shown in 21. 22) Divide a lid into 10 spokes. Trim and feather end of the spokes as shown. See number 22. 23) Divide a lid into 10 spokes. Trim ends of spokes and feather-cut edges as shown in number 23. 24) Divide a lid into 10 spokes. Trim the ends of the spokes as shown and feather-cut the edges of the spokes as shown. See 24. 25) Divide a lid into 10 spokes. Divide every other spoke in half, trim, shape and feather-cut edges as shown in 25.

Step Six: Filigreed medallions

Numbers 26, 27 and 28 are filigreed medallions. Make these as follows: 26) Divide a lid into 16 spokes and divide each spoke into five equal strips. Cut each strip to the center circle that defines the depth of cut. This divides the lid into 80 equal strips. Bend every other strip up at right angles to the lid. Bend every other strip of the strips bent up in toward the center of the lid. Curl the outer ring of strips to 1/2 inch height and the inner ring to 3/4 inch height. Curl the flat strips one turn. See number 26. 27) Mark a lid into 12 equal parts. Make a heavy line 1/8 inch to either side of every other line. This is to define the depth of cut of the strips. See Figure 1. Now cut down the other six lines dividing the can, thus giving six equal spokes, each one with a 1/4 inch ruled section in the middle. Figure 1 shows a series of lines, each one ending either at the inner circle or line in the center of the spoke. These lines represent the cuts that are made on each spoke. Follow this cut pattern. Start at the outside of edge of each spoke. Straighten the strips that are cut from each spoke before proceeding to the next, to avoid entanglement. Bend the three strips that are cut to the center circle up at right angles to the lid. Curl the center one of each group to 3/4 inch and the one to either side of center to 1/2 inch. Curl the rest of the strips as shown in 27, making one complete turn with the tool used, i.e., a single loop on the end of each strip. 28) Divide a lid into 8 equal spokes. Draw a line 1/8 inch from the right edge of each spoke on the side of the lid that is to be used as the face of the medallion. The inner circle on the lid and this line will define the depth of cuts that are to be made. Figure 2 shows the cut pattern for this medallion. Start at the left hand side of each spoke and cut into strips as indicated by the lines representing cuts in Figure 2. When cut, the strips will tend to curl up, and will naturally end in position to be curled. Curl the strips as shown in 28. The 4 or 5 longest strips are curled about 1-1/2 to 1-3/4 turns and the rest are curled 1 turn.

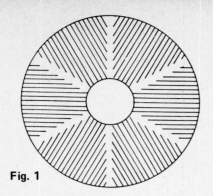

Fig. 1

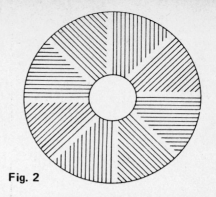

Fig. 2

Ornate Christmas Wreath

Materials:

1 - 18 inch diameter wire wreath form

125 - Can lids - assorted sizes - approximate

60 - Chenille stems - Same color as beads

1 - Piece of felt - 24 in. square. Color of beads

1 - Package of small size Christmas bead chain

1 - Package of medium size Christmas bead chain

1 Dozen 20 mm Christmas balls

1 - Wire coat hanger

36 - Small reflectors

12 - Medium reflectors

Bead wire

Bond cement

Step One: General description of wreath

The Christmas wreath is comprised of medallions of various sizes and shapes that are assembled or stacked to form decorative pieces. These pieces are decorated with Christmas ornaments and beads to add color and beauty to the finished wreath. The individual assemblies are then arranged and tied to the wreath frame to form the finished wreath. See the photograph.

It would be impossible to attempt to explain all the combinations of medallions that could be made into assemblies and the methods of decoration that could be used. Therefore, some of the specific medallions used in the illustrated wreath and the way that they are decorated will be explained and then, general procedure for assembling medallions so that the hobbyist can create his own decorative assemblies. There is no iron-clad rule used to make the wreath. The decoration and assembly are a matter of individual taste for the most part.

Fig. 1

Step Two: Medallion assemblies

The one thing that is common in all pieces used on the wreath, whether assemblies or individual medallions, is that three holes are punched, one in the center and one slightly to either side of center. A chenille stem that has been bent in half is inserted through the holes to either side of center. The stem is used to hold the unit on the wreath frame. When two medallions are stacked

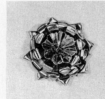

Fig. 2

Fig. 3

Fig. 4

Fig. 5

Fig. 6

Fig. 7

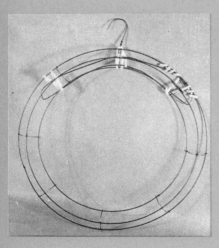

Fig. 8

Fig. 9

to form an assembly, they should be placed in the desired relation to one another and punched at the same time so that the holes will line up. When a reflector is used as the center of an assembly, it is held on by the wire or stem that holds the bead or ball type decoration in place. This wire is put through a hole punched in the center of the reflector, inserted through the center hole in the assembly and twisted around the chenille stem to fasten. Also, if a Christmas ball is used as a center decoration, the hole in the center must be large enough to accommodate the shank of the ball.

The above will not be repeated for each medallion assembly but should be referred to during assembly of units. Refer to page 151 for medallions used.

Some typical assemblies used on the illustrated wreath are: (1) Medallion 2 - background; number 18 - foreground. Position medallions as shown. String 8 medium size beads on bead wire. Position one between each pair of spokes on the foreground medallion and twist ends of bead wire to secure. String 4 medium size beads on bead wire, bring ends of wire together and twist. String 1 additional bead on loose ends and pull ends through center of group of 4 so that fifth bead is on top and in center. Put this assembly in center of piece, insert ends of wire through the center hole and twist around the chenille stem to fasten. See Fig. 1. (2) Medallion 12 - background; medallion 5 - foreground and medium size reflector in the center. The can lids are the same size. Place as shown, string 7 small Christmas beads on bead wire, insert through center holes of reflector and medallions and fasten around chenille stem. See Fig. 2. (3) Medallion 1 - background; number 17 - foreground. The can lids are the same size. Make "5 medium size bead center piece" as in number (1), above, and attach. See Fig. 3. (4) Medallion 3 - background; medallion 16 - foreground and small size reflector in the center. Arrange the medallions as shown, put the reflector in the center of medallion 16 and place a 20 mm Christmas ball in the center of the piece. See Fig. 4. (5) Medallion 4 - background; medallion 26 foreground. Arrange as shown, place a string of small Christmas beads around the upright strips of medallion 26, and a 20 mm Christmas ball in the center. See Fig. 5. (6) Medallion 27 with a 25 mm Christmas ball in the center. See Fig. 6. (7) Medallion 28 with 4 medium size Christmas beads in the center. See Fig. 7.

As can be seen from the examples above, the general method for making an assembly to place on the wreath is to use a "background", a "foreground", sometimes a reflector, and Christmas beads and balls for decoration. These are made with the background larger or the same size as the foreground lid. The above examples are just samples of the many medallion assemblies that can be used on the wreath.

Step Three: Assembly and finishing the wreath

A wire coat hanger should be bent to fit the shape of the wreath frame and wired to it. See Fig. 8. The finished medallions and medallion assemblies are fitted around the wreath frame and fastened to it with the chenille stems. The stem should be wound around the frame as shown. See Fig. 9. After several of the assemblies are on the frame it may be necessary to make special size assemblies to fit certain spots in the wreath. Also, you may find certain spots where it is difficult, if not impossible, to fit a medallion assembly. These places can best be filled with an appropriate sized reflector with bead ornamentation in the center.

When the wreath is filled with the medallion assemblies, place Bond cement on as many extremities of medallions as possible and on the wreath frame. Place the wreath in the center of the piece of felt, allow to dry and cut out as shown in photograph. An alternate way to attach felt is to stitch it on.

BREAD DOUGH
Artistry

.Jewelry
.Flowers
.Figures
.Filigree
.Salt Sculpture

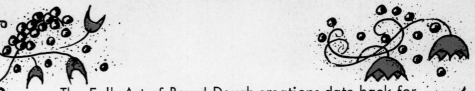

The Folk Art of Bread Dough creations date back for many centuries. Today's craftsmen are refining this old art by using Bread Dough - - a mixture of bits of bread mixed with white glue. The Dough is shaped into delicate flowers, jewelry, wall plaques, filigree trim, etc., a process which is a first cousin to ceramics and very similiar to working with air-drying clay. The results are true collector items with no two pieces alike.

We know that there will be as many "Bread Dough" recipes as there are recipes for "meat loaf" - - and this is good - - see basic recipe and variations below. We encourage you to try a "drop" of lemon, a "dash" of detergent whenever you wish to experiment - - that's being creative and having fun!

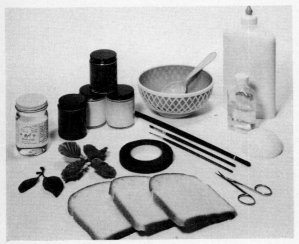

Materials And Supplies Needed

White bread
White glue
Poster paints - plain or
    fluorescent colors
Small brushes
Hazel's glaze
Glycerine
Floral tape

Wire for stems
Cuticle scissors
Scrap of white foam
Plastic or real leaves
Spoon for measuring
Bowl for mixing
Plastic bag for storing
Hand lotion

A commercial air-drying clay is available in craft departments which is excellent to use as a substitute for Bread Dough.

Basic Recipe For Bread Dough

3 SLICES WHITE BREAD
3 TABLESPOONS WHITE GLUE

Variations

3 drops glycerine
3 drops white shoe polish or
    3 drops Titanian white paint
3 drops lemon juice

Fig. A Tear the bread into small pieces.

Fig. B Mix ingredients together.

1. Remove crusts from bread.
2. Tear bread into small pieces.
Place pieces into a bowl - - see Fig. A.

3. Add glue plus one or all of the above variations. Mix together with fingers - - Fig. B. Knead until mixture no longer sticks to fingers and has a smooth texture.

This recipe is sufficient for most projects. Dough may be mixed ahead of time and stored in the refrigerator in a plastic bag --- it will remain useable for weeks.

1. After Bread Dough has been thoroughly mixed, dip a craft stick into jar of poster paint, "wipe" paint off on a piece of mixed Dough.*

2. "Wrap" Dough around the paint so that paint becomes distributed evenly throughout the Dough.

*It is best to break off 4 to 5 pieces of colorless Bread Dough, the approximate size of small walnuts. Add a different color to each: green, yellow, pink, blue, etc., or colors desired. To lighten color as it's being used, add a small amount of the colored Dough to some of the colorless Dough and mix together thoroughly. Place in a plastic bag to prevent drying.

"Tips And Tricks"

1. For ease in handling Bread Dough, put a small amount of glycerine or hand lotion on finger tips.

2. For minimum shrinkage, and to prevent larger items of the Bread Dough from possible cracking, coat them with a solution of equal parts water and white glue. Allow to dry, then repeat two more times. This will produce somewhat of a porcelain-like finish, but not as fragile as porcelain since the glue gives some elasticity to finished projects.

3. To thin edges of Dough when making leaves, etc., press Dough between thumbs and forefingers as in Fig. A at the right.

4. Set wired petals, leaves, etc., in a small piece of white foam to keep then in an upright position before assembling flower.

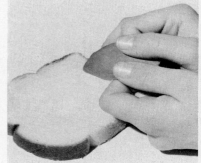

Fig. A Flatten edges of Dough between thumbs and forefingers.

5. Be sure to pull all of the stretch out of floral tape when wrapping flower stems or adding leaves and buds to stems.

6. When veining Bread Dough leaves, always press Dough against the underside of the real or plastic leaf - - this will produce the best veining.

7. Keep unused portions of mixed Bread Dough in a plastic bag to prevent air drying. This will keep several days in the refrigerator.

Finishes on Bread Dough

1. For a Bisque-like finish, allow completed project to dry without coating it with anything.

2. For a delicate ceramic-like finish, coat item 3 times with a solution of equal parts water and white glue, allowing it to dry in between each coat. After last coat, place in a 350 degree oven for 3 to 5 minutes. This produces a sheen on the project.

3. Finished projects may be painted with water base paints, acrylics or oils.

4. To give finished Bread Dough projects an opalescent look, spray with True Pearl.

5. Glaze may be brushed on to seal items and to give them a lasting shine.

Basic *Rose* Instructions

The delicate life-like Rose shown in the photograph at the left is approximately 1-1/2" in diameter. Made of a soft-hued pink Bread Dough, it is equally attractive when used in a miniature picture frame.

MATERIALS NEEDED:
Prepared Bread Dough - colorless
Poster paints - red, yellow, etc.,
    for rose, green for leaves
White glue
Green covered wire - 30 gauge
Floral tape - green

DIRECTIONS:
1. See page 157 "Coloring Bread Dough" and "Tips And Tricks".

    Color Dough as follows, (keep in mind inside petals of flowers are made from Dough a few shades darker than outside petals.)

        Red, yellow or pink for Rose
        Green for calyx and leaves

2. To form center of Rose, roll a small amount of Bread Dough into a "teardrop" shape. See Fig. A.

3. To make the stem, dip one end of a 5" length of 30 gauge covered wire into white glue, insert glue-coated end into the large end of "teardrop".

4. To make petals of Rose, roll Bread Dough into a small ball - - Fig. B. Flatten ball and shape in palm of hand or over knuckle of hand to form a cup shape. See Fig. C on opposite page. Smooth edges between fingers until they are very thin.

5. Use white glue to secure petals around "teardrop" center. Overlap at the bottom as shown in Fig. A. Pinch each petal at base.

6. Complete Rose as shown in Fig. A, by adding a second row of 3 petals, a third row of 3 or 5 petals. For a larger Rose add more rows of petals as desired. Use fingers to gently shape each petal. (Remember to add a touch of glue at bottom of Rose as each petal is added.)

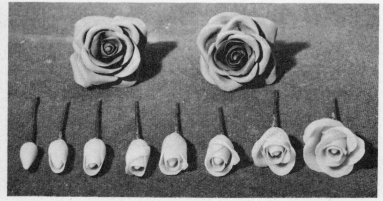

Fig. A. Step-by-step in adding petals to Rose.

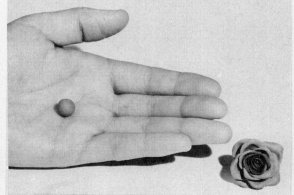

Fig. B. Roll Dough into a small ball.

7. To make calyx for rose, shape a green Bread Dough ball into a "teardrop". Divide the small end of teardrop into 5 equal parts by cutting Dough with cuticle scissors.

8. See Fig. D and use a toothpick to separate and flatten the parts of calyx.

9. Use cuticle scissors to cut lacy sections of calyx as in Fig. E.

10. Force Rose stem through the center of calyx, sliding calyx up the stem. Use white glue to glue calyx into place directly under the Rose.

NOTE: A small container may be made of Bread Dough similar to the one shown in the photograph on the opposite page.

Fig. C. Form Dough over knuckle to make petal.

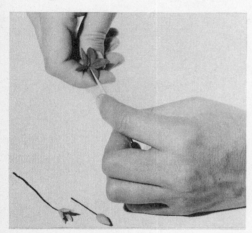

Fig. D Use a toothpick to flatten parts of calyx.

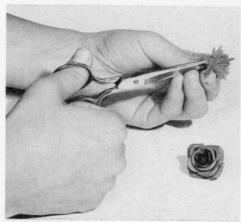

Fig. E Cut lacy sections with cuticle scissors.

1-3/4" diameter rose, leaves and even thorns are all made of Bread Dough. Stem is covered with brown floral tape.

Bread Dough roses are used to decorate the top of a gold leafed box - - ideal to use for dressing table to hold jewelry, etc.

1-1/2" diameter Poppy.

Basic *Poppy* Instructions

Before beginning, read General Instructions
on pages 156 and 157

MATERIALS NEEDED:

Prepared Bread Dough – colorless
Poster paints – yellow, orange, red and black
White glue
Square of nylon net, approximately 2"× 2"
Green covered wire – 30 gauge
Floral tape – green

DIRECTIONS:

1. **See page 157 for coloring Bread Dough.**
Color Dough as follows:
  Yellow for center
  Orange for center
  Black for fringe
  Orange or red for petals
  Green for leaves

2. Prepare center of flower using a 1/2" ball of yellow Bread Dough. Flatten to the size of a 25¢ piece. See pattern above. Place a small 1/4" orange ball onto center of the yellow – – see Fig. A. Wrap flattened yellow piece snugly around the orange ball, place in center of net square. Gather ends of net together, pushing Bread Dough through net slightly. Fig. B. Do not remove net from around the center, trim off only the ends.

Fig. A making center of Poppy

Fig. B pushing Dough through net

Fig. C roll Dough for fringe into a worm.

Fig. D cut fringe

Fig. E glue fringe around center

3. To add stem to center, dip one end of a 6" piece of covered 30 gauge wire into white glue. Insert glued end into bottom of center. Set aside to dry.

4. To make fringe around center, roll black Bread Dough into a plump 2" length "worm". Fig. C, opposite page. Flatten "worm" slightly and fringe one side by cutting with small cuticle scissors. Fig. D, opposite page. Use white glue to glue fringe into place around the flower center. Fig. E, opposite page.

5. To make oblong petal, place a flattened piece of Bread Dough between thumb and forefinger. Use thumb to push Dough forward to make petal curl. Fig. F. Curl one side only. Use white glue to secure curled petal around the fringe. Fig. G. Repeat making two more petals.

Fig.F curling petal

6. Proceed in the same manner for the leaves only curl both sides. Fig. H.

7. To add stem to leaf, dip a 3" piece of 30 gauge covered wire into white glue. Lay glue-coated end of wire onto the front center of leaf, pinching leaf slightly from the back so the Bread Dough will cover the wire in the front. Allow to dry.

8. To make the poppy bud, roll a greenish yellow piece of Bread Dough into a ball. Dip a 6" piece of 30 gauge covered wire into white glue - - insert glue-coated end of wire into bottom of ball.

Fig.G gluing petal

9. Use a razor blade to make impressions around top of the bud. Fig. I. Note: Bud may be made from colorless Bread Dough and painted with poster paints when dry.

10. To make calyx on bottom of bud, use cuticle scissors to cut 2 rows of 5 pointed stars. See Fig. J. Allow bud and flower to completely dry.

11. To assemble flower, wrap stem with floral tape, adding leaves and buds as you tape down the stem.

12. Completed flower may be placed in a small container that has been covered with bright colored burlap and decorated with quaint "homespun" braids or ribbons. See photograph on opposite page.

NOTE: For center variation, a plain ball of yellow or orange bread clay may be used. Make impressions around the top of the ball as for the bud in Step 9 - - then proceed as above.

Fig.H curling leaves

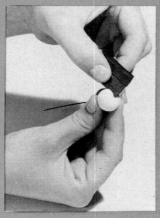

Fig.I make impressions around bud

Fig.J cutting calyx

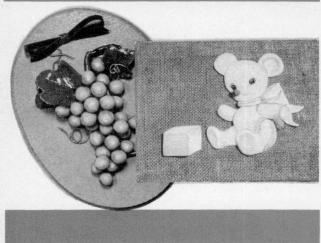

Bread Dough Plaques

These interesting plaques made of Bread Dough feature the use of wax cutters and simple line designs. Each plaque usually requires a recipe of Bread Dough containing approximately 6 to 8 slices of bread.

BASIC MATERIALS REQUIRED FOR EITHER TYPE PLAQUE

Bread Dough - colorless
Poster paints
White glue
Wax cutters or simple line design
Glaze
Brush
Background for mounting
Material or paint to cover background
X-acto knife
Wooden pointed tool
Plastic spoon

DIRECTIONS:

1. Select pattern or wax cutters to be used. Use poster paints to color Bread Dough
2. Use rolling pin and roll Dough between wax paper to approximately 1/8" to 3/16" thick.

. . . using tooling designs

3. Enlarge pattern if necessary - - lay on Dough and use x-acto knife to cut around entire pattern. For a 3-D effect, cut legs, arms, ears, etc., from more Dough and place on top of main body piece.

4. Smooth edges with fingers. Use a wooden pointed tool or plastic spoon to make muscle impressions, line ears, etc., while project is still moist.

5. Brush with a solution of 1/2 white glue and 1/2 water. Repeat 3 times, allowing project to dry in between coatings.

6. NOTE: Because of the size of these projects, dry thoroughly for several days before gluing onto background. Larger pieces such as these will shrink approximately 1/3 in size. To hasten drying, place project in a "slightly warm" oven.

7. When dry use white glue to glue into place on selected background.

8. Use poster paints to shade areas where necessary. Brush project with glaze, allow to dry.

. . . using wax cutters

Follow steps 1 and 2 above.

3. Use wax cutters to cut Dough. Smooth edges with fingers - - cup and shape each petal and leaf as desired. Make Dough ball grapes.

Follow steps 5 through 8 above.

NOTE: Place toothpicks in grapes for ease in handling. Larger petal edges should be propped up while moist so they will not "flop" while drying.

Decorative floral and fruit plaques shown at the right and above, are made from wax cutters - - the cute Teddy Bear above was made from a simple line design.

Framed Miniatures

Tiny Bread Dough flowers glued onto exquisite gilt frames make up these unique heirloom miniatures.

Photograph above shows the size of the miniatures in comparison with a tea cup. Including the frame, they measure from 2-3/4" to 7" in height.

Various kinds of Bread Dough flowers are used, along with 30 gauge covered wire glued into place for stems. See photographs.

Ideal to make and give as gifts, or for a decorative flair, group several together on a wall or hang three along a length of velvet ribbon.

Original Bread Dough Creations

The artistic Bread Dough creations shown on this page are but a few of the limitless items that can be made from a slice of bread, glue, a little paint and a flair of imagination!

Tiny "Potted" Plants

The colorful Pansies and Violets shown above are truly a masterpiece of artistic achievement. Authentically reproduced in color and size, the petals and leaves are veined and formed by pressing the colored Bread Dough onto the real leaves and petals. Use Poster paints and a tiny brush to add realistic shading.

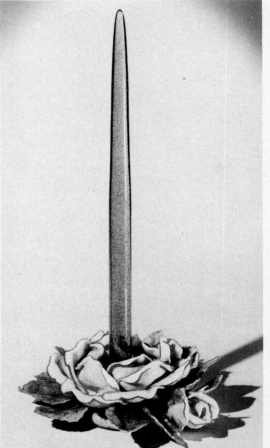

Candle Holder

Large 5-1/2" reproduction of a real live rose, the artistic Bread Dough candle holder rose shown at the left is made similar to the basic rose on page 4. A candle is pressed into the center before the Dough is dry to form a holder.

Wall Plaque

A walnut board forms the background for the wall plaque at the right. Borrow a tool from the family tool box - a 5/8" center punch to form blossom end of fruit and a "nail set" tool to form indentation on berries. Experiment with kitchen gadgets, etc., to see the types of impressions they leave on flexible Dough.

Bread Dough
Faces, Hands and Feet for
Collector Dolls

Collector Dolls are fun to make and Bread Dough may be used for heads, hands and feet, much the same as mache'. Interesting plastic molds are available for many faces while hands and feet may be shaped by hand.

Directions for heads made in plastic molds: See step-by-step photos for using plastic molds on page 176.

1. Lightly dust mold with cornstarch and then pack with flesh-colored Bread Dough.
2. Turn out face onto a piece of mylar or aluminum foil.
3. Cut excess Dough from around edges with an x-acto knife. Use fingers to round off any sharp edges.
4. A pointed wooden tool or toothpick is ideal to use to accentuate any details on the face.
5. Coat molded head 3 times with a solution of equal parts water and white glue. Allow glue to dry between each coating.
6. Place on foil in a 350 degree oven for 3 minutes only. Remove and cool.
7. The back flat portion of head may be attached to a half white foam ball by gluing together or with short lengths of 18 gauge covered wire. Dip ends of wire into white glue and then force one end carefully into back of head and other end into foam. Fig. A. Add another length of wire into bottom of foam ball for neck or stem with which to attach head to a body of your choice.
8. Cover back of foam head and around face with "hair" - - (curly angel hair, string hair, crochet cotton hair, etc.)

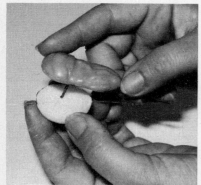

Fig. A attaching head to half white foam ball.

Hands, feet, or shoes:

These may be molded free-hand using a flesh-colored Bread Dough. Coat with a glue and water solution as in step 5 above. Add heavy covered wire with which to attach hands, etc., to the body. Fig. B.

Fig. B wire stem inserted into Bread Dough hand

Painting:

Use poster paints and a tiny brush to paint desired details and shading onto head, hands, feet or shoes, and then coat with a clear glaze for permanency.

Holiday Poinsettia

Patterns on next page

MATERIALS NEEDED:

Bread Dough
Poster Paints – red, green and yellow
White glue
30 gauge uncovered wire
Floral tape – green

SUPPLIES NEEDED:

Small brush
Cuticle scissors
Scrap of white foam
Plastic Poinsettia leaf and petal

NOTE: Read General Instructions for mixing and coloring Bread Dough on pages 156 and 157. See opposite page for Petal Patterns and amount of Dough required.

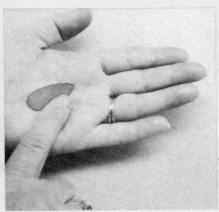

1 MAKING RED PETALS
Use pea-size ball of Dough. Roll into cone and flatten in palm of hand. Thin edges between fingers.

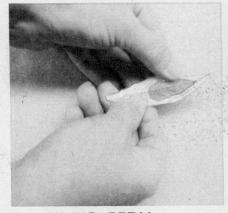

2 VEINING PETAL
Select plastic petal, larger than Dough petal. Center flattened Dough against vein on underside of plastic petal.

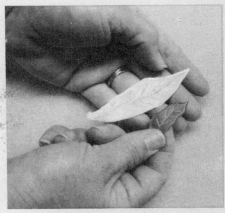

3 REMOVE VEINED PETAL
Pull Dough petal from plastic petal. Same plastic petal is used for veining all petals.

4 WIRING PETAL
Place glue coated end of wire 1/2" along top of main vein. Gently pinch backside so that Dough covers wire and shapes petal.

5 SET PETALS ASIDE
Stick wire ends of petals into a scrap of white foam while making other petals.

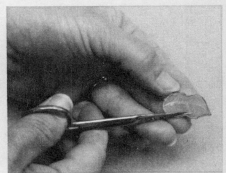

6 GREEN LEAVES
Make leaves same as petals. Shape with cuticle scissors, vein and add stem wire.

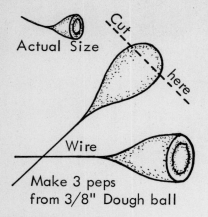

Actual Size

Cut here

Wire

Make 3 peps from 3/8" Dough ball

7 MAKING PEPS

Shape 3 small yellow cones. Cut large end off – use a toothpick to "etch" a circle on flat cut end – insert glue coated wire into pointed end.

8 ASSEMBLY

First, arrange smallest petals around peps, wire to hold. Add other petals. Lastly, tape leaves into place.

9 SHADING

Use poster paints to shade:
Petals – Dark Red
Leaves – Dark Green
Pep Center – Dark Red

Size and Color of Dough

Color 3/8" diameter Dough ball yellow for Peps.

Color 1-1/4" diameter Dough ball Red for Petals.

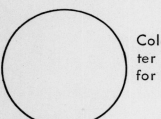
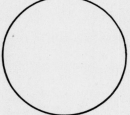

Color 1-1/4" diameter Dough ball Green for Leaves.

Suggested Petal Patterns

Need 7 of each size Petal for Poinsettia

Truly a masterpiece of artistic expression, the colorful Poinsettia shown in the photograph above measures approximately 4-1/2" in diameter. Surprisingly easy to make by following the step-by-step instructions, the Bread Dough artisan will be gratified with the outstanding results.

The delicate beauty of this life-like flower was enhanced by wiring it to the black wrought iron candle sconce - - a decorative conversation piece wherever used.

Bread Dough Jewelry

These exquisite heirloom jewelry pieces are fashioned of tiny Bread Dough flowers that measure 1/8" to 1/4" in diameter.

MATERIALS NEEDED:

Bread Dough - colorless
Poster paints - red, pink, etc., for flowers
Green for leaves
Covered 22 gauge wire
White glue
Pendant background
Earring background
Earring clips
Jump rings and chain

Plastic domes
Cuticle scissors
X-acto knife
Green thread for stems
Tweezers and toothpick
Jewelry pliers
Velvet for background

DIRECTIONS:

1. Color Dough as desired for flowers. Form ball of Dough approximately 1/8" in diameter.

2. Flatten ball between thumb and forefinger to form petal.

3. To form a bud, dip one end of a 2" length of 22 gauge wire into white glue. Place wire end onto edge of flattened petal and roll petal around wire. Slip bud off the wire and set aside to dry.

4. Repeat same procedure as above to begin rose. Do not remove from wire. Place glue along bottom edge of second flattened petal - - place bottom of petal into position just above bottom of bud. Flair tip of petal outward and press bottom securely around bud. Repeat with another petal on opposite side of bud to make a small rose.

5. To make larger roses, place 7 petals around bud. Slip rose off wire - - if glue has dried, cut wire off at bottom of rose.

6. Make as many flowers as desired to fill backgrounds, see photos on opposite page. Pendants usually have from 3 to 7 flowers and earrings have 2 to 3 flowers.

7. To make leaves, flatten a small 1/8" ball of green Dough between fingers. Use cuticle scissors to cut the leaf shape. Use an x-acto knife or edge of scissors to press veining impressions onto leaf.

8. Cut cardboard pattern the size to fit the back of pendant setting. Cut velvet background from cardboard pattern. Fig. A. Use white glue to glue velvet and background into place.

9. Use white glue to glue flowers into place onto velvet background - - see Fig. B. Use tweezers and toothpick for ease in handling tiny flowers if necessary.

Fig. A Cut velvet for background

Fig. B Glue flowers into place

Fig. C Trim plastic domes

10. To make flower stems, cut 1/4" to 1/2" lengths of green thread -- glue into place under flowers.

11. Glue leaves into place at random -- see photograph at the right. Allow to dry thoroughly.

12. Use an x-acto knife or cuticle scissors to trim around plastic domes -- Fig. C on opposite page. Place small amount of white glue along edge of plastic dome and fit into place on background.

13. Attach jump rings, chain and earring clips.

Tiny Bread Dough jewelry items that are easily made and are surprisingly durable and sturdy.

Bread Dough "Blossom Clusters"

"Swinging Blossom Cluster" earrings are ideal for color accent.

Minute blossom clusters for earrings or mini-potted plants are so easy to make and are surprisingly light in weight!

#1 in photo -- form tiny closed blossoms as shown. To make open blossoms, use cuticle scissors and "cross-cut" the top of closed bud. Use a toothpick to separate and spread blossom open. Brush on a "touch" of yellow poster paint for a center to open blossom.

#2 in photo -- use a toothpick to pierce holes in a 1" white foam ball. Dip ends of blossoms and buds into white glue and push into holes.

#3 in photo -- completely cover ball with blossoms. Attach earring clips, etc., as desired with white glue.

#4 in photo -- add stem wire to tiny blossoms and put in a mini-pot made of Bread Dough.

169

Fantasy Flower and Little Chick

This Bread Dough flower features the use of parts of a nylon hose to make the unusual center and stamens. Petals are of yellow Dough, center is white and stamens are of orange Dough.

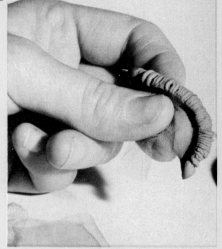

Fantasy Flowers measures 4" in height.

1. Push Dough through nylon hose for the fine center. Insert wire for a stem.

2. Push Dough through a run in the nylon hose for stamens. Trim off excess hose.

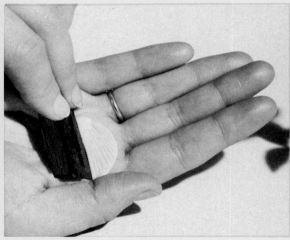

3. After shaping, use a razor blade to line petals. Glue into place around the stamens.

4. Press Dough onto a real leaf to make impressions - - trim to desired size. Add stem wire - - use floral tape to tape leaves onto stem.

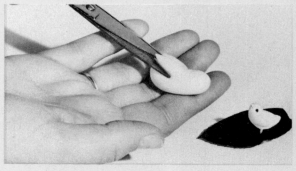

5. Shape "chick" as above from a ball of yellow Dough the size of a large pea.

6. Use tips of scissors to cut wings on each side of chick - - smooth dough out.

Ecuadorian Bread Dough Art

Bread Dough creations from Ecuador, such as the ones shown in the photographs below, date back for centuries. Originally, they were prepared as offerings of food on All Saint's Day, and the tradition is still carried on. No two pieces are ever alike because each tiny piece is hand shaped and carefully assembled. Natural dyes are used to create the artistic projects as true-to-life as possible. After completion items are always lacquered to preserve the vibrant colors and designs.

HOLIDAY TREE DECORATIONS

"DOUGH" PIN-ONS

Ideas . . .

A gold leaf frame enhances the beauty of the elegant spray of Bread Dough flowers glued onto a white china dinner plate -- so beautiful to display or give as a gift!

"Tropical Jungle" scene above, features the use of a wire armature for branches. Dough is formed over wire -- coat Dough with water and glue solution 3 times to prevent cracking. Paint as desired.

Bread Dough Figures

These clever 2-1/4" to 2-1/2" size Bread Dough animals are but a sample of the many novel designs that may be made. Fun to use as teenage pins, tree ornaments, party favors, etc.

MATERIALS NEEDED:
Bread Dough - colorless
Poster Paint - black
Fluorescent Poster Paints - yellow, red, orange
　　　magenta, chartreuse, green and cerise
X-acto knife
White glue
Hazel's Glaze
Toothpick
Optional: Pin backs or string for hanging

Owl

DIRECTIONS:
1. Trace and cut master cardboard pattern for body from pattern given below.
2. Flatten Bread Dough to approximately 3/16". Place pattern on Dough and use an x-acto knife to cut around pattern.
3. Use fingers and gently smooth cut edges of Bread Dough and shape body so it is slightly rounded.
4. For eyes, beak, feet and trim, color small amounts of Bread Dough, black, yellow, red, orange, magenta, chartreuse, green and cerise.
5. Eye of owl is made of 4 colors of Bread Dough "stacked" and glued together as follows:

　　　●　　3/8" tiny flattened ball of black Dough

　　　●　　5/16" tiny flattened ball of yellow Dough

　　　●　　3/16" minute flattened ball of red Dough

　　　●　　1/8" pin-point ball of black Dough

NOTE: Use a toothpick for ease in picking up and placing tiny pieces of Bread Dough.
6. See patterns and shape beak and feet from red Bread Dough. Glue into place on owl as shown in photograph. Use a toothpick to make small impressions on beak.

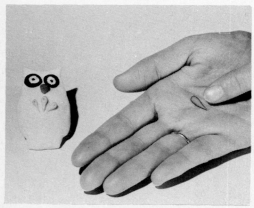

Fig. A Roll teardrop of Dough

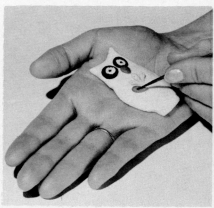

Fig. B Use toothpick for lifting

BEAK PATTERN

FEET PATTERN

7. To make colorful trim, roll small balls of different colors Bread Dough into "teardrops" measuring from 1/4" to 3/8" in length. See Fig. A. on opposite page. Glue trim onto owl as in photograph. (A toothpick is used to help pick up "teardrops", leaving a small impression on large end of trim. See Fig. B on opposite page.)

8. Secure pin back or string onto owl before body is completely dry. Place a dab of glue onto a flattened 1/4" ball of colorless Bread Dough – – immediately place over the pin back or end of string and smooth into place. Fig. C at the right. Allow owl to thoroughly dry.

9. For a finished look, brush front side of owl with a coat of Hazel's Glaze.

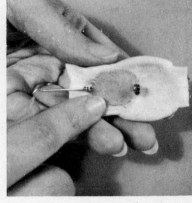

Fig. C Secure pin back into place

Turtle

DIRECTIONS:

1. Trace and cut master cardboard pattern for body of turtle from pattern given below.

2. To make body, flatten a ball of colorless Bread Dough to approximately 3/16" thick. Place pattern on Dough and use an x-acto knife to cut around pattern.

3. Place body in palm of hand and use finger to cup as in Fig. D. Smooth and shape cut edges of body with fingers.

4. To make head and tail, roll a piece of Dough approximately 3/16" thick, 2-1/2" in length and tapered as in the pattern Fig. E. Glue into place on underside of body. See Fig. F.

5. To make legs, roll a piece of Dough approximately 3/16" thick by 3-1/2" in length. Cut length in half and shape as in Fig. G. See Fig. F photograph for placement and use white glue to glue into place.

6. Decorate shell of turtle with colorful Bread Dough trim as shown in design pattern above. Trim eyes, feet and head.

7. To attach string or pin back see step 8 above. Allow turtle to dry before brushing on Hazel's Glaze.

PATTERN FOR DESIGN

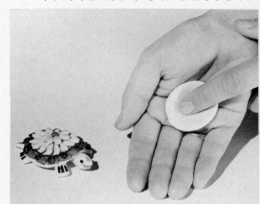

Fig. D Shape body in palm of hand

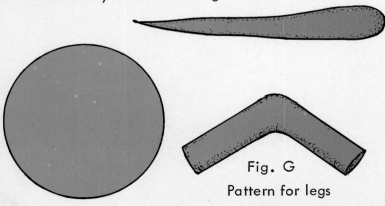

Fig. G

Pattern for legs

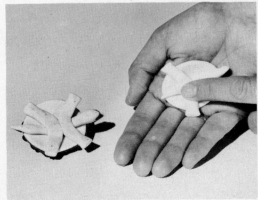

Fig. F Glue body and legs into place

Salt Sculpturing

A wonderful world of fun and surprises awaits youngsters and grownups alike when working with this "magic modeling goop".

BASIC RECIPE
4 cups unsifted flour
1 cup salt - (plain or iodized)
1-1/2 cups water
NOTE: Recipe should not be doubled or halved.
Alternate recipe given below

1. Mix the ingredients together with your fingers. If the mixture is too stiff, add a little more water. When thoroughly mixed, knead the mixture for 4 to 5 minutes.

2. Use balls of mixture to form figures, hanging medallions, flowers, animals, etc., (see photographs). Many kinds of kitchen utensils come in handy for shaping and imprinting figures.

3. Hooks to hang the finished projects may be made from hair pins, wire, paper clips, etc., and should be inserted in figures before they are dry. Metal buttons are sometimes used as trim on projects and are baked right into the figure.

4. To bake, pre-heat oven to 350 degrees. Bake each piece for one hour. For thicker or larger pieces bake longer. Use a toothpick at a joint to test for doneness. If it is soft, leave it in the oven a while longer.

5. Occasionally a piece will brown unevenly, but this will not detract from its appearance when it is painted. Remove from the oven to cool before painting.

6. To decorate, use poster paints or felt tip markers - - and use your imagination!

7. Always coat completed pieces with a clear glaze or a fixative. The salt in the mixture draws moisture from the air and might eventually soften pieces.

THINGS TO REMEMBER
For all flowers, roll mixture very thin between sheets of wax paper - - use a knife to cut desired shape. Other flowers may be cut with wax cutters or cookie cutters.

To give projects a textured appearance, as for angel robes, etc., roll the mixture between terry cloth towels.

For very large projects such as wreaths, use chicken wire as a base to "build" on.

Don't try to make upright pieces unless you have an armature.

Remember, DO NOT allow anyone to eat mixture or completed projects.

ALTERNATE RECIPE
2 cup salt
2/3 cup water
1 cup loose cornstarch
1/2 cup cold water

Mix salt and 2/3 cup water in a saucepan. Stir until mixture is well heated - - 3 to 4 minutes. Remove from heat and add cornstarch which has been mixed with 1/2 cup cold water. Stir quickly. Mixture should be the consistency of stiff dough. Place over low heat and stir approximately one minute until mixture forms a smooth pliable mass.

Makes approximately 1-3/4 lbs., and can be kept indefinitely if wrapped in plastic or foil. No refrigeration necessary.

Original Salt Sculpturing Projects

Many interesting as well as decorative projects may be made from this easy to use Salt Sculpturing mixture. Shown in the photographs are but a few of the numerous ideas from which to start working. Many of the smaller hanging projects are used for Christmas tree ornaments, while larger pieces are used as centerpieces, etc. Jewelry takes on an unusual look of originality when crafted from the Salt Sculpturing mix. Use a toothpick to pierce through the soft "bead" - - when dry, combine them with wooden beads when stringing.

Bread Dough Filigree For Decorating

The most decorative of all box decor is the molded Bread Dough Filigree, so easily made in the numerous styles of plastic molds available in hobby and craft departments.

MATERIALS NEEDED:

Wooden box - size of your choice
Sandpaper
Spray paint - color of your choice
Plastic filigree mold of your choice
Cornstarch for mold release
Bread Dough - colorless
Poster paints - if pre-coloring Bread Dough
White glue
Liquid Gold Leaf
Mylar
Hazel's Glaze

DIRECTIONS:

1. Prepare wooden box by sanding lightly to achieve a smooth surface.

2. Spray paint box with a color of your choice.

3. Determine placement of filigree on box.

4. Prepare parts to be used in plastic mold by dusting it lightly with cornstarch for a mold release.

5. Press Bread Dough firmly into the mold, making sure to fill all areas of the mold.

6. Immediately lift a corner of the molded filigree and gently pull it away from the mold.